# Principles of Genetic Toxicology

## Second Edition

# Principles of Genetic Toxicology

## Second Edition

**David Brusick**

*Hazleton Laboratories*
*Kensington, Maryland*

SPRINGER SCIENCE+BUSINESS MEDIA, LLC

Library of Congress Cataloging in Publication Data

Brusick, David.
Principles of genetic toxicology.

Bibliography: p.
Includes index.
1. Genetic toxicology. I. Title. [DNLM: 1. Environmental Pollutants. 2. Mutagens.
3. Mutation. QH 465.C5 B912p]
RA1224.3.B78 1987 575.2'9 87-17169
ISBN 978-1-4899-1982-3 ISBN 978-1-4899-1980-9 (eBook)
DOI 10.1007/978-1-4899-1980-9


Originally published by Plenum Press, New York in 1987
Softcover reprint of the hardcover 2nd edition 1987

**This book is dedicated to**

**Dr. Alexander Hollaender**
**(1899–1986)**

# Foreword

The field of genetic toxicology has gone through remarkable development in the seven years since the appearance of the first edition of *Principles of Genetic Toxicology*.

One branch of toxicology research, chemical mutagenesis, has been elucidated and expanded as a result of increased effort, testing, and the sharing of data. This expansion has occurred not only in the industrialized countries, but also in countries that are comparatively less advanced in scientific implementation. These developing countries have taken advantage of the basic practical methods that were so well described in the first edition of this work. It is significant to note how many centers have been established throughout the world and are now studying the basic concepts and applying them to practical problems such as the detection of genetic effects caused by exposure to chemicals. In fact, there are now toxicology training centers in twelve countries. Genetic toxicology, in addition to being investigated as a science unto itself, has been taught to people in the applied fields so that these techniques may be put to use in solving other biological problems.

For these reasons, it is most useful to have an update of the basic methods and their development. Dr. Brusick should be congratulated for doing such an excellent job of assembling a text that will be worthwhile to any researcher who is interested in the principles of genetic toxicology.

Alexander Hollaender
*Council for Research Planning in Biological Sciences, Inc.*
*Washington, D.C. 20036*

# Preface

*The present generation is only a caretaker of the human genome of future generations.*
MALLING AND VALCOVIC (1978)

The above quotation from Malling and Valcovic was used in the first edition of this work. It is so accurate and states so clearly the task placed before genetic toxicologists that it is used again to set the tone of this edition.

Since the first edition of this book was published in 1980, genetic toxicology has encountered numerous challenges and has participated in several scientific breakthroughs, which will affect the course of this discipline for many years to come. Yet at the same time some of the fundamental problems that faced genetic toxicologists in 1980 remain in 1987 with no appearance of forward progress.

The objectives of this edition remain essentially the same as those of the first edition. It strives to provide a synthesis and perspective on genetic toxicology that might in some way help practitioners and individuals just entering this area of toxicology to make real progress toward solving some of the difficult problems, and to help assure that our generation will be a responsible caretaker of the gene pool.

David J. Brusick

*Kensington, Maryland*

# Contents

*Chapter 9*



*Chapter 10*

CHAPTER 1

# Origins of Genetic Toxicology

## INTRODUCTION

The field of toxicology deals with effects of agents on living systems, with the purpose of defining human health effects. It is an applied science, which draws on data and methodology from a multitude of basic sciences, such as physiology, pharmacology, the study of metabolism, ethology, genetics, embryology, chemistry, and statistics.

Over the past 20 years, toxicology has provided the primary source of data on health effects for chemical safety evaluations on existing and new products. Without this information, many potentially hazardous chemicals would have been identified only through human experience. Accidental exposure to chemicals, such as the methyl isocyanate accident in Bophal, India in 1984, and exposure to chemical warfare agents has provided significant data to toxicologists. Approved experimentation on the terminally ill has also played an important role in developing basic information regarding expression of toxic phenomena in humans. While important data have been obtained in such instances, these do not constitute an acceptable means for the evaluation of new chemicals prior to their introduction into the environment.

Animal models are generally used as human surrogates in toxicologic evaluations. Animal models are bioassays that involve the use of a species believed to exhibit a toxic response similar to that of humans under exposure conditions relevant to human exposures. Mammals are the most common test organisms employed. The choice of a mammalian species depends on two factors: its suitability as a model for the human experience, that is, its close anatomic and physiologic resemblance to humans; and economic considerations, such as availability and cost of the animal.

For the assessment of acute, subchronic, and chronic toxicity, rodent species are considered suitable, but for specific types of toxicity only certain mammals respond in a manner similar to humans. This is particularly true for psychopharmacology and studies of toxicity associated with nutrition or aging.[1,2] Compromises have been made in some areas, such as the study of reproduction and teratology. Rodents are often used in routine testing, but in special cases primate species may be required.

Numerous other subspecialties of toxicology must rely on model systems likely to be suboptimal because suitable surrogates of the human effects are not available. Examples are the study of neurotoxicity, skin sensitization, dermal absorption and toxicity, and possibly certain types of carcinogenesis.[3–5] Toxicity assessment is clearly an inexact science, and extrapolation of effects observed in animal models exposed to chemicals under laboratory conditions to humans is tenuous. Animal testing involves the use of genetically homogeneous animals exposed under carefully regulated laboratory environments. Neither of these situations duplicates the typical human exposure.

Reliable estimates of human risk also depend upon an understanding of the cellular and molecular mechanisms that underlie toxic phenomena expressed at the whole organism level, and on adequate methods of quantitating both the true absorbed dose of a chemical (as well as the amount reaching the critical target site) and the magnitude of the toxic response. Dose is most often equated with exposure, that is, the environmental concentration surrounding the test organism, which seldom represents the dose delivered to the critical target.

The study of animal toxicology may be divided into many subspecialties, each representing a separate field of study. Some of the subspecialties include metabolism/pharmacokinetics, genetic toxicology, reproduction/teratology, immunotoxicology, and behavioral toxicology.

One unifying superstructure among all subdivisions of toxicology is the duration of the study. Most studies fall under one of three categories: acute, subchronic, or chronic. The duration of each type of study depends on the test organism, but in general for small rodents, acute studies run from 1 to 15 days, subchronic studies from 15 to 120 days, and chronic studies from 12 months to the lifetime of the animal. Table 1.1 summarizes some of the typical toxic endpoints measured and the duration of the study associated with them. Within each category of toxicologic assessment the types of tests vary considerably in their form. Historically, the largest proportion of toxicology tests have been descriptive in nature. Elaborate compilations of animal observations, body weights, and histopathology have been used to identify and describe toxic phenomena produced by a chemical exposure. During the past 10 years, interest has increased in analytical bioassays, which attempt to study the mechanistic relationship

TABLE 1.1
Typical Toxic Endpoints or Signs Observed in the Major Temporal Subdivisions of Animal Toxicology

| Acute | Subchronic[a] | Chronic[b] |
|---|---|---|
| Lethality | Changes in reproductive capacity | Nonreversible tissue degeneration |
| Irritation | Teratologic effects | Carcinogenesis |
| Necrosis | Reversible tissue degeneration | Changes in life expectancy |
| Changes in normal homeostasic parameters | Dietary deficiencies | |
| Neurologic effects | Behavioral changes | |

[a] Subchronic effects may also include those listed under Acute.
[b] Chronic effects may also include those listed under Acute and Subchronic.

between exposure to a chemical or class of chemicals and the toxic syndrome expressed. This interest has led in great part to the expansion of toxicology specialization.

## THE ROLE OF GENETIC TOXICOLOGY IN HEALTH EFFECTS TESTING

Genetic toxicology, as a subspecialty of toxicology, identifies and analyzes the action of agents with toxicity directed toward the hereditary components of living systems. Some toxicants damage the genetic complex at concentrations also producing acute nonspecific cytotoxicity and death; the primary objective of genetic toxicology, however, is to detect and analyze the hazard potential of those agents that are highly specific for interactions with nucleic acids and produce alterations in genetic elements at subtoxic concentrations. Compound exposures for genetic toxicology studies range from acute to chronic, thus falling into the three major temporal subdivisions of toxicologic testing.

Agents that produce alterations in the nucleic acids and associated components at subtoxic exposure levels, resulting in modified hereditary characteristics or DNA inactivation, are classified as genotoxic. Genotoxic substances usually have common chemical or physical properties that facilitate interaction with nucleic acids. In fact, the universality of the target molecule is the key to the discipline of genetic toxicology. Derivation of the term genotoxic is found in a publication by Ehrenberg *et al.*[6] A report of the International Commission for Protection Against Environmental Mutagens and Carcinogens (ICPEMC) provided a more detailed definition for the term genotoxic and emphasized that categorization of a chemical as genotoxic is not an *a priori* indication of a health hazard.[7] The term is a general descriptor meant to distinguish chemicals that have an affinity for direct DNA interactions from those that do not.

## HISTORY OF GENETIC TOXICOLOGY

Genetic toxicology evolved from the initial studies of gene mutability demonstrated first by Muller in 1927 using radiation,[8] followed almost 20 years later by Auerbach *et al.* using chemicals.[9] Both of these investigators conducted their studies using submammalian species, but within the next 20 years, genetic changes in animals induced by radiation and chemicals were demonstrated by Cattanach[10] and Russell[11] at Oak Ridge. This work created the awareness that some of the "hereditary" diseases observed in human populations might be environmental in origin.[12,13] Proof in the early 1940s that deoxyribonucleic acid (DNA) is the hereditary material, and the subsequent elucidation of its primary, secondary, and tertiary structures by Watson and Crick in the early 1950s, opened up new avenues of research into the mechanisms of mutagenesis.

The period of time from 1953 to 1968 might be considered the first "Golden Era" of molecular genetics. During this time, much of the basic information was developed regarding DNA structure and replication, the genetic code, mechanisms of protein synthesis, and DNA repair processes (Table 1.2). Cell biologists and biochemists and microbiologists reigned over this Golden Era, and several received Nobel Prizes for their contributions. The foundation established by the scientists in the first "Golden Era" has led to a second burst of information and technology associated with directed manipulation of genetic elements, often referred

TABLE 1.2
Major Advances in Molecular Genetics

| Phenomenon | Year | Reference |
|---|---|---|
| DNA constitutes the hereditary material | 1944 | 14 |
| Elucidation of DNA structure | 1953 | 15 |
| Mechanism of protein synthesis—the central dogma hypothesis | 1957 | 16 |
| Demonstration of semiconservative replication of the DNA molecule | 1958 | 17 |
| Operon model of gene regulation | 1961 | 18 |
| Elucidation of the genetic code for protein | 1964–1967 | 19 |
| Enzymatic synthesis of DNA *in vitro* | 1967 | 20 |
| Gene cloning (transpecies) | 1977 | 22 |

to as genetic engineering. This new era started in 1978 with the reports of gene transfer and expression between unrelated organisms.[21,22]

Genetic toxicology was recognized as a discipline in 1969 when the Environmental Mutagen Society was founded under the leadership of Dr. Alexander Hollaender and several other geneticists who were concerned about the potential genetic impact associated with the proliferation of man-made chemicals in the environment. The concept of genetic toxicology was clearly consistent with the intense concern for environmental protection that prevailed at the time.

Occurring simultaneously with the growing concern by geneticists over environmental mutagens were the reports from several independent groups of investigators showing a correlative relationship between mammalian carcinogens and mutagens.[23–27] Earlier attempts to support such a correlation experimentally failed because of the limitations inherent in the genetic assays available.[28] The concept of mutagenic carcinogens was revitalized, however, following the introduction of procarcinogen activation using host-mediated[29] and *in vitro* microsome activation systems[30] developed from 1969 through 1971. The fulfillment of this relationship was the identification of mammalian cancer genes (oncogenes) and the demonstrated presence of mutations in the activated forms of these genes.[31]

Thus, genetic toxicology has played a dual role in safety evaluation programs. One function is the implementation of testing and risk assessment methods to define the impact of genotoxic agents found in the environment and whose presence may alter the integrity of the human gene pool. The second function is the application of genetic methodologies to the detection and mechanistic understanding of carcinogenic chemicals. In this latter regard, genetic toxicology has been applied as a front-line screen for potential carcinogens.

Genetic toxicology has continued to grow as a specialty area of toxicology, with significant influence on overall chemical safety evaluation.

## TECHNOLOGY TRANSFER AND APPLIED GENETICS

The application of genetic methodology to human problems occurred prior to the recent concern over genotoxicity. The breeding of plants and animals for improved food production is derived from early experimentation in genetics. This applied science is an international, multibillion-dollar business involving the development of new varieties of plants and animals for recreational as well as agricultural purposes.

Technology transfer, the application of basic science methodology to the solution of practical problems, has been essential to the development of genetic toxicology. Many of the methods employed to detect genotoxic substances were initially developed for other purposes. The Ames *Salmonella*/microsome assay, for example, evolved from early biochemical investigations of histidine biosynthesis in *Salmonella typhimurium*.[32] The UDS (unscheduled DNA synthesis) assays used to measure stimulation of DNA repair processes were originally part of research efforts directed toward the elucidation of enzymes present in DNA repair pathways.

Toxicology has traditionally been an animal testing science relying primarily on the detection of qualitative and quantitative changes in physiology, behavior, homeostatic processes, and lethality. From such observations, a toxicity profile of a substance is generated. This profile is then used to assess the probable human health consequences following exposure to the substance, and it often forms the basis for establishing permissible environmental levels of the agent.

As a consequence of numerous scientific and regulatory pressures, interest in understanding the molecular mechanisms for all toxic phenomena has increased. This new orientation of toxicologic investigation has been identified as molecular toxicology. Molecular toxicologists seek to define mechanistic similarities and differences between experimental systems and humans. Genetic toxicology, behavioral toxicology, immunotoxicology, teratology, and oncogenesis are disciplines of toxicology in which molecular toxicologists have been most active. Stimulating the development of molecular toxicology are the increased financial and moral costs associated with animal testing. These costs continue to rise at the same time the need for evaluating the toxicity of more chemicals is growing. Thus, searches for rapid, economical, nonanimal predictive tests based on an understanding of molecular processes will continue to expand. Technology transfer from the basic sciences is expected to produce a minor revolution in the field of toxicology during the next decade.

## THE COMPONENTS OF GENETIC TOXICOLOGY

Technology transfer has produced a remarkable turnover in genetic toxicology assay systems during the past several years. New test systems are evaluated for their suitability, and those with promise are often incorporated into the array of tests employed in this science. Unfortunately, as new test systems have been added, few have been retired. Thus, genetic toxicology is cluttered with tests of limited value and relevance for real genotoxic assessment. For example, tests for phage induction (induct test)[33] and biphenyl hydroxylation[34] appeared promising in early phases of validation, but they have exhibited low levels of compound discrimination when tested with a wider range of chemical classes and are not now generally considered reliable indicators of genetic hazard. It is possible that these tests will have value in restricted situations. Another test, the host-mediated assay, originally described by Gabridge and Legator,[29] is still used, but has undergone modifications that have significantly enhanced its utility compared with the original methodology.[35,36] The original purpose of host-mediated techniques was to provide a means for metabolic activation of promutagens, but the demonstration by Malling in 1971 that *in vitro* microsome metabolic systems could be coupled with microbial mutagenicity testing essentially replaced host-mediated methods as screening tests.[30]

Some types of test methods that had lost popularity with the introduction of S9-supplemented microbial and mammalian cell culture tests are being rediscovered. Extensive research involving *Drosophila melanogaster* has renewed interest in this classic genetic test organism. *Drosophila* has many advantages, including its own metabolic activation system and the ability to screen for both gene and chromosome effects.[37] A shortened somatic cell version of the *Drosophila* assay has recently been proposed for screening.[38]

New tests using mammals offer the possibility of measuring a broad array of genetic endpoints *in vivo*. These tests encompass both somatic and germ cell target sites.[39]

Thus, the entire array of tests used to detect genotoxic compounds covers a broad range of techniques, including some assays of limited use, such as the silkworm oocyte assay[40] and mutant induction in *Tradescantia* stamen hairs,[41] as well as more common assays, such as the dominant lethal test derived from traditional animal toxicology techniques. Only a few tests are presently included in routine screening programs found in toxicology testing laboratories.

The route of new tests from the experimental to the applied stages generally follows a similar pattern (Figure 1.1). The first step is development and confirmation of a technique, which is often derived from fundamental research. This step includes the verification of the nature of the endpoint (i.e., the endpoint must be a true genetic alteration, not a phenomenon resembling a genetic event). The standard methodology and protocol are also established during the development and verification stage. The development process usually employs only well-defined model chemicals. The second step is one of validation. The overall utility of an assay rests on its ability to detect genetic activity in a wide range of chemical classes. Once a relatively standardized protocol is established, an assay is subjected to vigorous evaluation in multiple laboratories using coded samples. The final step is the application of the assays in "real-life" testing with a uniform protocol. The U. S. Environmental Protection

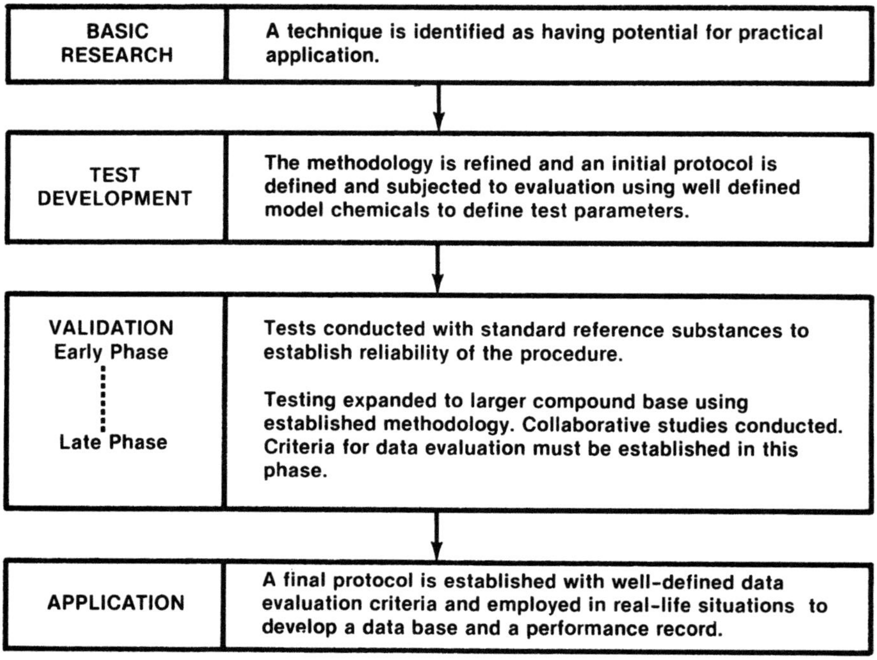


FIGURE 1.1. Steps involved in technology transfer for new assays.

Agency (EPA) under its Gene-Tox program has developed an approach to determine the maturity (level of development) for a large number of genetic bioassays.[42] These criteria include the number of chemicals and chemical classes represented in the published literature, the level of protocol standardization, the number of laboratories using the method, and the number of chemicals evaluated with the test per year.

As in other areas of toxicology, a secondary type of technology transfer is gaining momentum in genetic toxicology. Computer-based data handling and reporting systems as well as on-line, computer-monitored equipment are greatly improving the turnaround time and overall capacity of testing laboratories. Data management is rapidly becoming one of the most serious generalized problems facing toxicology testing. The reasons involve not only the need to generate rapidly raw data into tabular form, but also regulatory requirements for long-term data storage and retrieval.

Genetic toxicology is likely to undergo several significant evolutionary steps in the near future. Application of new methods developed in the biotechnology revolution will add new dimensions to genetic testing (see Chapter 9), and the number of tests will be reduced and refined as data bases grow larger and those tests with serious limitations are identified. Sophisticated instrumentation and computer-based data collection, analysis, and reporting will also increase the accuracy and data handling capacity of this area of toxicology.

## REFERENCES


1. Burnet, F. M.: *Intrinsic Mutagenesis: A Genetic Approach to Aging*, Medical and Technical Publishing, Lancaster, England, 1974.
2. Hanin, I., and Usdin, E. (eds.): *Animal Models in Psychiatry and Neurology*, Pergamon Press, New York, 1978.
3. Brusick, D. J.: The value and significance of carcinogenic, mutagenic, and teratogenic tests. In *Cutaneous Toxicity* (V. A. Drill and P. Lazar, eds.), Academic Press, New York, pp. 189–201, 1977.
4. Shubik, P.: Identification of environmental carcinogens. Animal test models. In *Carcinogens: Identification of Mechanisms of Action* (A. C. Griffin and C. R. Shaw, eds.), Raven Press, New York, pp. 37–47, 1979.
5. Wilson, J. G.: *Environment and Birth Defects*, Academic Press, New York, Chapter 8, 1973.
6. Ehrenberg, L., Brookes, P., Druckrey, H., Lagerlof, B., Litwin, J., and Williams, G.: The relation of cancer induction and genetic damage. In *Evaluation of Genetic Risks of Environmental Chemicals*, Report of Group 3, Ambio Special Report No. 3, Royal Swedish Academy of Sciences, Universitetsforlaget, 1973.
7. ICPEMC: Committee 1 Final Report. Screening strategy for chemicals that are potential germ-cell mutagens in mammals. *Mutat. Res.* **114:**117–177, 1983.
8. Muller, H. J.: Artificial transmutation of the gene. *Science* **66:**84–87, 1927.
9. Auerbach, C., Robson, J. M., and Carr, J. G.: The chemical production of mutations. *Science* **105:**243, 1947.
10. Cattanach, B. M.: Chemically induced mutations in mice. *Mutat. Res.* **3:**346–353, 1966.
11. Russell, W. L.: X-ray induced mutations in mice. *Cold Spring Harbor Symp. Quant. Biol.* **16:**327–330, 1951.
12. Bridges, G. A., Bochkov, N. P., and Jansen, J. D.: Genetic monitoring of human populations accidentally exposed to a suspected mutagenic chemical. International Commission for Protection against Environmental Mutagens and Carcinogens, Publication No. 1, *Mutat. Res.* **64:**57–60, 1979.
13. Russell, W. L.: The role of mammals in the future of chemical mutagenesis research. *Arch. Toxicol.* **38:**141–147, 1977.
14. Avery, O. T., MacLeod, C. M., and McCarty, M.: Studied on the chemical nature of the substance inducing transformation of pneumococcal types. *J. Exp. Med.* **79:**137–158, 1944. Reprinted in *Classic Papers in Genetics* (J. A. Peters, ed.), Prentice-Hall, Englewood Cliffs, New Jersey, pp. 173–192, 1959; and *Papers on Bacterial Genetics* (E. A. Adelberg, ed.), Little, Brown, Boston, pp. 147–168, 1960.
15. Watson, J. D., and Crick, F. H. C.: The structure of DNA. In *Papers on Genetics: A Book of Readings* (L. Levine, ed.), C. V. Mosby, St. Louis, pp. 11–21, 1971.
16. Crick, F. H. C.: On protein synthesis. *Symp. Soc. Exp. Biol.* **12:**138–163, 1958.
17. Meselson, M. S., and Stahl, F. W.: The replication of DNA in *Escherichia coli*. *Proc. Natl. Acad. Sci. USA* **44:**671–682, 1958.
18. Jacob, F., and Monod, J.: On the regulation of gene activity. *Cold Spring Harbor Symp. Quant. Biol.* **26:**193, 1961.

19. Nirenberg, M., Caskey, T., Marshall, R., Brimacombe, R., Kellogg, D., Doctor, B., Hatfield, D., Levin, J., Rottman, F., Pestka, S., Wilcox, M., and Anderson, F.: The RNA code and protein synthesis. In *Papers on Genetics: A Book of Readings* (L. Levine, ed.), C. V. Mosby, St. Louis, pp. 34–52, 1971.
20. Goulian, M., Kornberg, K., and Sinsheimer, R. L.: Enzymatic synthesis of DNA, XXIV. Synthesis of infectious phage OX174 DNA. In *Molecular Biology of DNA and RNA: An Analysis of Research Papers* (I. D. Raacke, ed.), C. V. Mosby, St. Louis, pp. 112–119, 1971.
21. Guarente, L., Lauer, G., Roberts, J. M., and Ptaskne, M.: Improved methods for maximizing expression of a cloned gene: A bacterium that synthesizes rabbit β-globin. *Cell* **20**:543–553, 1980.
22. Itabura, K., Hirose, T., Crea, R., Riggs, A. D., Heyneker, H. L., Bolivar, F., and Boyer, H. W.: Expression in *Escherichia coli* of a chemically synthesized gene for the hormone somatostatin. *Science* **198**:1056–1063, 1977.
23. Ames, B. N.: Identifying environmental chemicals causing mutations and cancer. *Science* **204**:587–593, 1979.
24. Bridges, B. A.: Short-term screening tests for carcinogens. *Nature* **261**:195–200, 1976.
25. Brusick, D. J.: The role of short-term testing in carcinogen detection. *Chemosphere* **5**:403–417, 1978.
26. McCann, J., Choi, E., Yamasaki, E., and Ames, B. N.: Detection of carcinogens as mutagens in the *Salmonella*/microsome test: Assay of 300 chemicals. *Proc. Natl. Acad. Sci. USA* **72**(12):5135–5139, 1975.
27. Sugimura, T., Sato, S., Nagao, M., Yahagi, T., Matsushima, T., Seino, Y., Takeuchi, M., and Kawachi, T.: Overlapping of carcinogens and mutagens. In *Fundamentals in Cancer Prevention* (P. N. Magee, T. Matsushima, T. Sugimura, and S. Takayama, eds.), University of Tokyo Press, Tokyo, and University Park Press, Baltimore, Maryland, pp. 191–215, 1976.
28. Brusick, D. J.: *In vitro* mutagenesis assays as predictors of chemical carcinogenesis in mammals. In *Toxicology Annual*, Vol. 2 (C. L. Winek, ed.), Marcel Dekker, New York, pp. 79–105, 1977.
29. Gabridge, M. G., and Legator, M. S.: A host-mediated assay for the detection of mutagenic compounds. *Proc. Soc. Exp. Biol. Med.* **130**:831, 1969.
30. Malling, H. V.: Dimethylnitrosamine: Formation of mutagenic compounds by interaction with mouse liver microsomes. *Mutat. Res.* **13**:425, 1971.
31. Reddy, E. P., Reynolds, R. K., Santos, E., and Barbacid, M.: A point mutation is responsible for the acquisition of transforming properties by the T24 human bladder carcinoma oncogene. *Nature* **300**:149–152, 1982.
32. Whitfield, H. J., Martin, R. G., and Ames, B. N.: Classification of aminotransferase (C gene) mutants in the histidine operon. *J. Mol. Biol.* **21**:335, 1966.
33. Moreau, P., Bailone, A., and Devoret, R.: Prophage λ induction in *Escherichia coli*, K12 *env* A, *uVr* B: A highly sensitive test for potential carcinogens. *Proc. Natl. Acad. Sci. USA* **73**:3700–3704, 1976.
34. McPherson, F., Bridges, J. W., and Parke, D. V.: *In vitro* enhancement of hepatic microsomal biphenyl 2-hydroxylation by carcinogens. *Nature* **252**:488–489, 1974.
35. Arni, P., Mantel, T., Deparade, E., and Muller, D.: Intrasanguine host-mediated assay with *Salmonella typhimurium*. *Mutat. Res.* **45**(3):291–307, 1977.
36. Durston, W. E., and Ames, B. N.: A simple method for the detection of mutagens in urine: Studies with the carcinogen 2-acetylaminofluorene. *Proc. Natl. Acad. Sci. USA* **71**:737–741, 1974.
37. Vogel, E., and Sobels, F. H.: The function of *Drosophila* in genetic toxicology testing. In *Chemical Mutagens: Principles and Methods for Their Detection*, Vol. 4 (A. Hollaender, ed.), Plenum Press, New York, Chapter 38, pp. 93–142, 1976.

38. Graf, U., Wurgler, F. E., Katz, A. J., Frei, H., Juon, H., Hall, C. B., and Kale, P. G.: Somatic mutation and recombination test in *Drosophila melanogaster*. *Environ. Mutagen.* **6**:153–188, 1984.
39. Malling, H. V., and Valcovic, L. R.: New approaches to detecting gene mutations in mammals. In *Advances in Modern Toxicology*, Vol. 4 (G. Flamm and M. Mehlman, eds.), Halsted Press, New York, Chapter 8, 1978.
40. Tazima, Y., and Onimaku, K.: Results of mutagenicity testing for some nitrofuran derivatives in a sensitive test system with silkworm oocytes. Japanese Environmental Mutagen Society, 2nd Annual Meeting, Abstract 16, 1976.
41. Underbrink, A. G., Schairer, L. A., and Sparrow, A. H.: *Trandescantia* stamen hairs: A radiobiological test system applicable to chemical mutagenesis. In *Chemical Mutagens: Principles and Methods for Their Detection*, Vol. 3 (A. Hollaender, ed.), Plenum Press, New York, Chapter 30, p. 171, 1976.
42. Brusick, D., and Auletta, A.: Developmental status of bioassays in genetic toxicology. *Mutat. Res.* **153**:1–10, 1985.

CHAPTER 2

# Fundamentals of Genetic Toxicity

## INTRODUCTION

The purpose of this chapter is to describe the genetic background and terminology essential for an understanding of genetic toxicology. Since the types of molecular lesions that chemicals induce in DNA and the resultant genetic damage are intimately tied to the structure and function of this molecule, it is essential to appreciate the basic structure of DNA and understand how it operates.

Because of the unique position held by DNA in storing and processing cellular information, an intrinsic capacity for self-repair appears to have evolved simultaneously with the environmental adaptation of this molecule. This repair process is believed to influence the kinetics of mutation induction by preventing many induced lesions from becoming fixed as permanent damage.

The first portion of this chapter presents some of the fundamentals of the science of genetics. The second section is devoted to a presentation of the categories and mechanisms of DNA alterations that generate genotoxic effects in cells.

## BASIC GENETICS FOR TOXICOLOGISTS

### Gene Structure

DNA is the macromolecule from which all characteristics of life ultimately derive (Figure 2.1). The informational molecules of all living systems, with the exception of some viruses that use ribonucleic acid

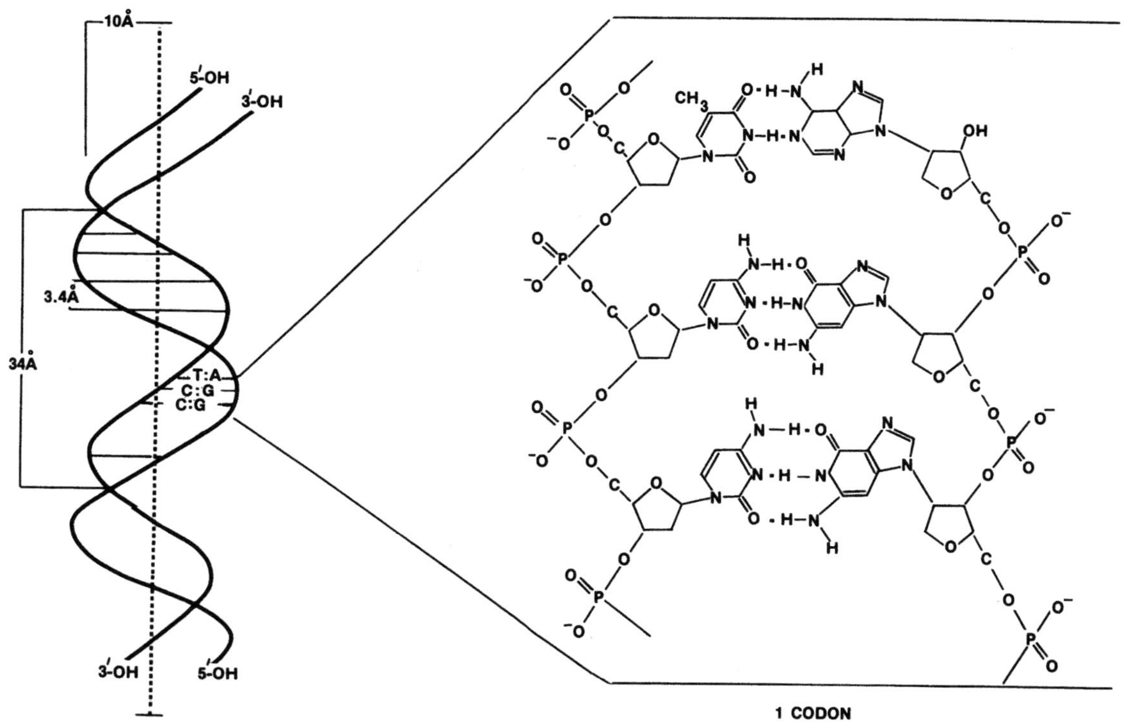


FIGURE 2.1. The basic structure of DNA. An enlarged codon consisting of three base pairs is shown. This enlarged view illustrates the base configuration and hydrogen bonding relationships for DNA. Hydrogen bonding involves weak interactions between N, O, and H atoms and allows the two strands to separate from each other over short regions or during replication. One codon specifies an amino acid.

TABLE 2.1
Basic Biochemical Characteristics of All Double-Stranded DNA

1. DNA consists of two different purines (guanine, adenine) and two different pyrimidines (thymine and cytosine)
2. A nucleotide pair consists of one purine and one pyrimidine [adenine/thymine (AT) or guanine/cytosine (GC)]
3. Nucleotide pairs are connected into a double-helix molecule by sugar–phosphate backbone linkages and hydrogen bonding (Figure 2.1)
4. The AT base pair is held by two hydrogen bonds, while the GC is held by three
5. The distance between each base pair in a molecule is 3.4 Å, producing ten nucleotide pairs per turn of the DNA helix
6. The number of adenine molecules must equal the number of thymine molecules in a DNA molecule; the same relationship exists for guanine and cytosine molecules; the ratio of AT to GC base pairs, however, may vary in DNA from species to species
7. The two strands of the double helix are complementary and antiparallel with respect to the polarity of the two sugar–phosphate backbones, one strand being 3′–5′ and the other being 5′–3′ with respect to the terminal OH group on the ribose sugar
8. DNA replicates by a semiconservative method in which the two strands separate and each is used as a template for the synthesis of a new complementary strand
9. The rate of DNA nucleotide polymerization during replication is approximately 600 nucleotides/sec; the helix must unwind to form templates at a rate of 3600 rpm to accommodate this replication rate
10. The DNA content of cells is variable ($1.8 \times 10^9$ daltons for *Escherichia coli* to $1.9 \times 10^{11}$ daltons for human cells)

(RNA), are composed of DNA. Some of the characteristic features of DNA molecules are listed in Table 2.1. The fact that the DNA of all living organisms, from bacteria to humans, has these characteristics emphasizes the antiquity of its origin and the immortality of those genes that pertain to the cellular housekeeping processes of DNA synthesis, replication, and repair. The mechanisms of information storage and gene expression are extremely similar in all organisms.

The simplest complete functional unit in a DNA molecule is termed a "gene." Most of what is known about the structure and operation of genes has been acquired from studies with bacteria or bacteriophages.[1–3] The differences between the genes of prokaryotic organisms (bacteria) and eukaryotic organisms (plant and animal cells) center primarily on their number, location on the respective chromosomal entities, and mechanisms of gene regulation. In prokaryotic cells there is a single chromosomal entity with little or no differentiation along the DNA molecule as far as function is concerned. Eukaryotic cells, on the other hand, have DNA with nonfunctional, repeated sequences of some genes; these cells also have regions of noncoding DNA, called introns, between coding sequences called exons. The function of repeated DNA sequences and intron regions is not known. The nucleotide composition and the mechanisms by which information encoded in a gene are transformed into gene products appear to be universal. This has been confirmed by the recent developments in recombinant DNA genetic engineering, in which genes continue to function properly after having been transplanted from human cells to bacterial cells or from bacterial to plant cells.[4,5] Universality in all phases of gene decoding is fundamental to the success of this type of genetic research.

## Gene Function

With only minor exceptions, it is assumed that gene function at the molecular level parallels DNA structural similarity and is identical in all organisms. The following characteristics represent the basics of gene function.

1. A gene is composed of a segment of a DNA molecule containing sufficient information to produce a functional product. The products of

TABLE 2.2
Dictionary of the Genetic Code for Proteins[a]

| Codon | Amino acid | Codon | Amino acid | Codon | Amino acid | Codon | Amino acid |
|---|---|---|---|---|---|---|---|
| UUU | Phe | UCU | Ser | UGU | Cys | UAU | Tyr |
| UUC | Phe | UCC | Ser | UGC | Cys | UAC | Tyr |
| UUA | Leu | UCA | Ser | UGA | None | UAA | Ochre |
| UUG | Leu | UCG | Ser | UGG | Trp | UAG | Amber |
| CUU | Leu | CCU | Pro | CGU | Arg | CAU | His |
| CUC | Leu | CCC | Pro | CGC | Arg | CAC | His |
| CUA | Leu | CCA | Pro | CGA | Arg | CAA | Glx |
| CUG | Leu | CCG | Pro | CGG | Arg | CAG | Glx |
| AUU | Ile | ACU | Thr | AGU | Ser | AAU | Asx |
| AUC | Ile | ACC | Thr | AGC | Ser | AAC | Asx |
| AUA | Ile | ACA | Thr | AGA | Arg | AAA | Lys |
| AUG | Met | ACG | Thr | AGG | Arg | AAG | Lys |
| GUU | Val | GCU | Ala | CGU | Gly | GAU | Asp |
| GUC | Val | GCC | Ala | GGC | Gly | GAC | Asp |
| GUA | Val | GCA | Ala | GGA | Gly | GAA | Glu |
| GUG | Val | GCG | Ala | GGG | Gly | GAG | Glu |

[a] U substitutes for T in RNA. The DNA code sequence would be complementary to each RNA code [e.g., (DNA sequence/CAG) → (RNA sequence/GUC) → (amino acid/valine)]. Codon words read from 5′-OH end (left) to 3′-OH end (right). Codons UGA, UAA, and UAG are all terminating sequences and do not code for any amino acids.

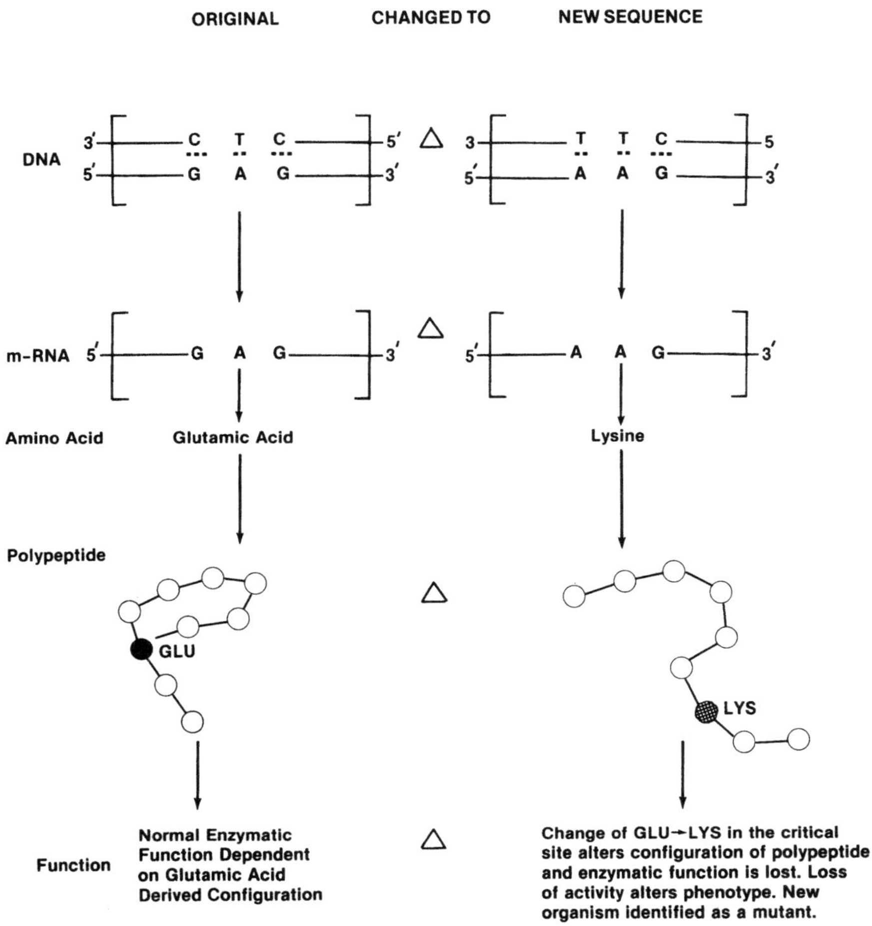


FIGURE 2.2. Gene–protein relationships in mutation induction. A change in the base sequence of one codon may produce an amino acid replacement in the gene product that results in a nonfunctional polypeptide.

almost all genes are polypeptides (amino acid heteropolymers). The function of these polypeptides can be structural, enzymatic, or regulatory. All visible and functional qualities of an organism, defined as the phenotype, are dependent upon the accurate structure and expression of these gene products.

2. The information required to specify a single amino acid is contained in a triplet of base pairs called a codon. The sequence of the bases in the codons is important for translation of information. Table 2.2 gives

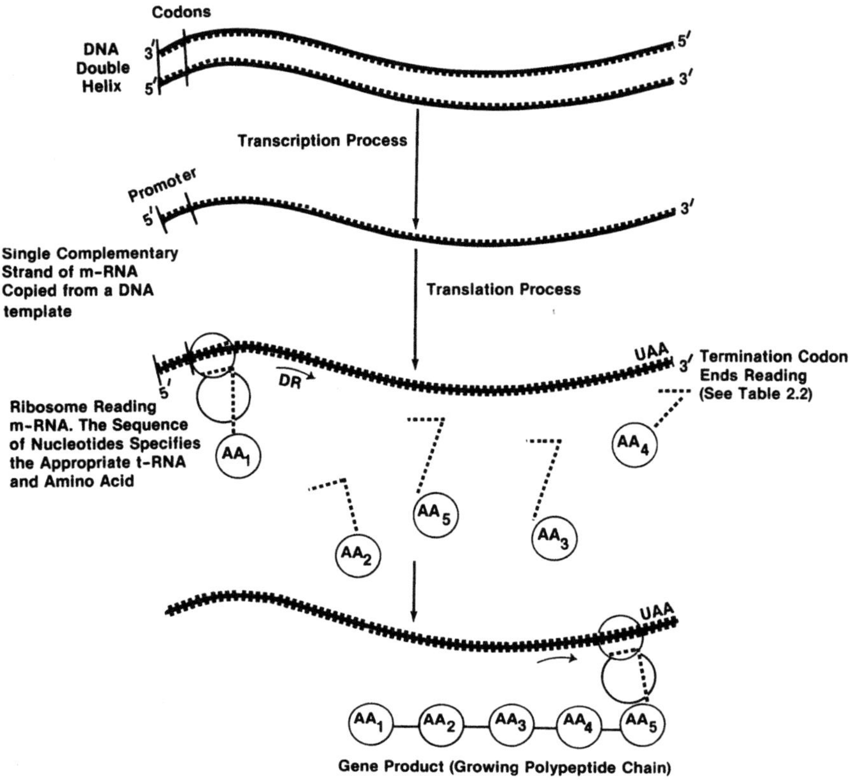


FIGURE 2.3. Essential steps in prokaryotic protein synthesis. Protein synthesis involves two major actions: (a) decoding DNA information into RNA information (transcription) and (b) decoding RNA information into polypeptides (translation). The reference to codons in the top portion of the figure encompasses more than three base pairs and hence more than one codon. mRNA, messenger RNA; tRNA, transfer RNA; AA, amino acid; DR, direction of reading; $AA_{1-5}$, amino acids.

the codon dictionary for the naturally occurring amino acids used to synthesize protein. More than one codon can specify a given amino acid. For example, 5′ CGU 3′, 5′ CGA 3′, and 5′ AAG 3′ all code for the amino acid arginine. The sequence of codons in a gene specifies the sequence of amino acids in the polypeptide and thus its ultimate role in cellular processes.

3. Changes in the composition of the base pair of a codon or changes in the codon composition of a gene can result in a gene product that will not function. This may eventually be expressed by cell or organism death, or in an altered form of the cell or organism (Figure 2.2). Thus, these changes form the basis for genetic alterations, or “mutations.”

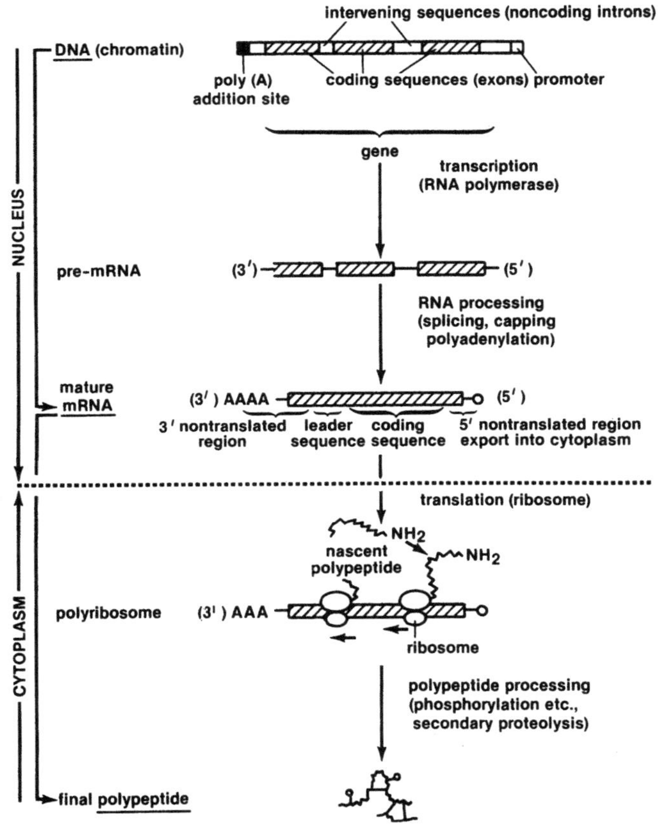


FIGURE 2.4. Eukaryotic gene expression. The process of gene expression in eukaryotic cells follows a pattern similar to that in prokaryotic cells except that portions of the process occur in the nucleus and portions occur external to the nucleus. The pre-mRNA step is not found in prokaryotic cells, which do not have intron sequences. The mature mRNA is transported across the nuclear membrane to the cytoplasmic area where ribosomes are located. Translation of the mature mRNA is then similar in both prokaryotic and eukaryotic organisms. (Courtesy of P. Lohman.)

4. The production of a polypeptide gene product involves two types of RNA: messenger RNA (mRNA), a short-lived copy of the gene being expressed; and transfer RNA (tRNA), which contributes to both the transportation and coupling of amino acids into the polypeptide gene product. This latter process also involves an RNA/protein structure called a ribosome. The basic process identified in prokaryotic (bacterial) cells is illustrated in Figure 2.3. In eukaryotic cells the process of gene expression follows a similar pattern, but intron regions must be removed from the

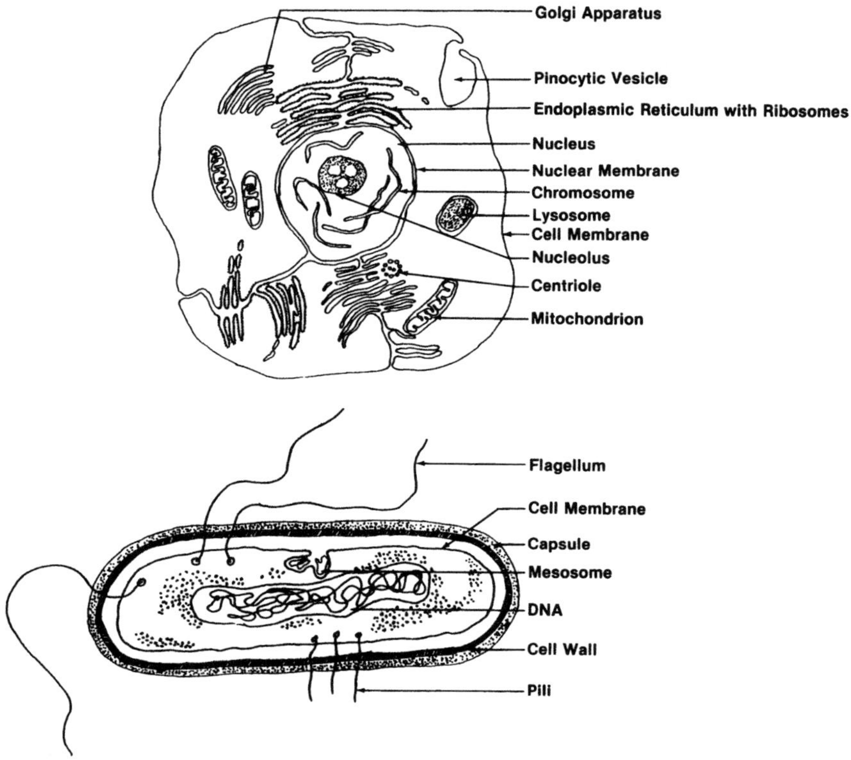


FIGURE 2.5. Prokaryotic and eukaryotic cell types. Substantial differences are observed in subcellular organization. These illustrations are generalizations; many variations of each cell type exist. (Illustrations courtesy of H. Lebowitz.)

mRNA before translation occurs (Figure 2.4). Enzymes located in the nucleus of the cell excise intron regions and splice the coding sequences back together. The resulting mRNA is then transported to ribosomes outside the nucleus for translation.

5. The initiation and termination of the transcription of a gene into a polypeptide are regulated by a separate set of regulatory (promoter) genes. Most regulatory genes respond to chemical or temporal cues such that only those genes needed at a given time are expressed. The remaining genes are in an inactive state. The processes of gene activation and inactivation are believed to be critical to cellular differentiation in multicellular organisms.

Nucleic acid organization at the cellular level, as well as gene regulation, differ in prokaryotic and eukaryotic cell types (Figure 2.5). For

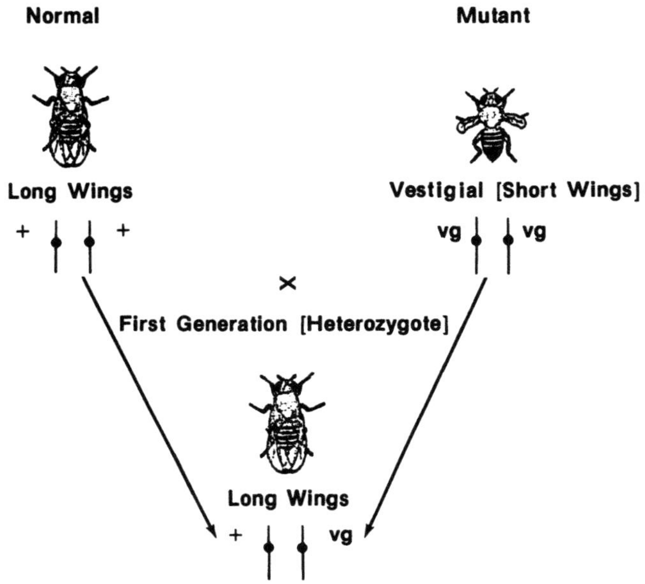


FIGURE 2.6. Dominant and recessive traits. Long-wing trait is dominant and will be expressed either as +/+ or +/*vg*. Short-wing trait is recessive and will be expressed only when the genotype is *vg*/*vg*. This figure also illustrates the relationship between genotype (composition of the gene pair) and phenotype (visable appearance of the organism). For example, two different genotypes (+/+ and +/*vg*) can both exhibit the same phenotype.

example, DNA location and packaging are distinctly different in bacterial (prokaryotic) and mammalian (eukaryotic) somatic cells. These differences will more likely influence gene expression mechanisms than mutation induction mechanisms. Once a genotoxic agent reaches DNA (prokaryotic or eukaryotic), mutation induction processes are probably identical. The following series of statements represents the basics of nucleic acid organization at the cellular level.

1. DNA may be packaged with proteins into chromatin material and organized into chromosomes, or it may be present as an uncomplexed DNA helix as in a prokaryotic cell.

2. Chromosomes may be located unrestrained with the prokaryotic cell or within a nuclear membrane in the eukaryotic cell.

3. Chromosomes may be found as single copies per cell (haploid), or in multiple copies per cell, generally two (diploid). In haploid cells, all functional genes present in the cell can be expressed. In diploid cells, one gene may be dominant over the other so that only the dominant gene of each functional pair is expressed. The nonexpressed gene of the pair is

TABLE 2.3
Characteristics of DNA in Prokaryotic and Eukaryotic Cell Types

| Prokaryotic | Eukaryotic |
|---|---|
| Primarily haploid | Primarily diploid |
| DNA uncomplexed | DNA complexed with proteins forming chromosomes |
| DNA nonlocalized in the cell cytoplasm | DNA localized primarily within the nucleus of the cell |
| No morphologic stages in DNA replication | DNA replication described by mitotic cycle consisting of specific cytologic stages |
| DNA often found as a closed circle | DNA found in linear chromosomes |
| Replication not associated with cellular organelles | Replication and separation of chromosome associated with cellular organelles called centrioles |
| All genes encoded in the DNA are functional | Repetitive, nonfunctional gene sequences are common |
| Spacer sequences have not been identified | Noncoding spacer sequences identified as introns occur along the DNA molecule |

called recessive. Functional recessive genes are expressed only when both copies of the recessive type are present (Figure 2.6). Some cell types in mammals are found in ploidies other than diploid. Liver cells, for example, are often found to be tetraploid, and gametes are, of course, haploid.

Alternate forms of the preceding characteristics are functions of whether or not the organism is prokaryotic or eukaryotic in nature. Table 2.3 summarizes some of the general traits of prokaryotic and eukaryotic cells as they pertain to the organization of genetic material. It is essential to delineate the similarities and differences between prokaryotes and eukaryotes, since both cell types are routinely employed in genetic testing and some of the differences appear to be critical in specific testing situations, such as target cell metabolism and ability to express aneuploidy.

## THE CELL CYCLE AND CHROMOSOME MECHANICS IN SOMATIC AND GERM CELLS

### Somatic versus Germ Cells

Mammalian organisms consist of several distinct cell types that have acquired specific functions during differentiation. All of the cell types in the mature organism originated from one of the three embryonic cell types (mesodermal, entodermal, or ectodermal). Histopathologic diagnosis of toxic effects relies upon examination of the morphology and histochemistry of differentiated cell types. Issues relevant to genetic toxicology focus more on the hereditary status of the target cell.

Somatic cells constitute the major portion of the mammalian organism. The genomes of most somatic cells are diploid. Damage to somatic cells is not transmissible to subsequent generations and may have toxicologic consequences only to the exposed cell population. Virtually all *in vitro* assays measure damage in somatic cells. Germ cells are those cells that undergo meiotic division to haploid gametes. Damage to germ cells has the potential for transmission to the next generation. A description of the relative sensitivities of these two cell types to genotoxic agents is given in Chapter 5.

## The Chromosome

The DNA molecule of a bacterium is not generally referred to as a chromosome, and cell cycle and chromosome mechanics are not particularly relevant with respect to the types of genetic lesions induced in these organisms. Most test systems in genetic toxicology, however, employ eukaryotic organisms (fungi, insects, plants, mammals), and cell cycle processes become important in conducting the assay and interpreting the results.

Chromosomes consist of DNA containing several hundred to thousands of genes, depending upon the species. Complexed with this DNA is a group of basic proteins called histones and some nonhistone proteins.[6] Histones attach preferentially to specific sites on the chromosomes and are considered important in the regulation of gene function, since they appear to inhibit the expression (transcription) of the genes to which they are complexed.[7] The union of DNA and proteins provides the basis for the cytologically visible features of chromosomes. This feature is important in cytogenetic assays that rely on microscopic examination of chromosome number and structure. A typical condensed metaphase chromosome illustrating each of these features is shown in Figure 2.7.

Each metaphase chromosome shows a primary constriction called the centromere. The constriction may occur near the middle of the chromosome (metacentric) or near one end (subterminal or acrocentric). The position of the centromere is useful in distinguishing chromosomes, since its location for a given chromosome is constant in a species. Functionally, the centromere is believed to be the attachment site for the spindle apparatus connecting the centriole to the chromosome. This attachment is necessary to separate chromatids during cell division (see Figure 2.8).

A secondary constriction in one or more chromosomes within a set

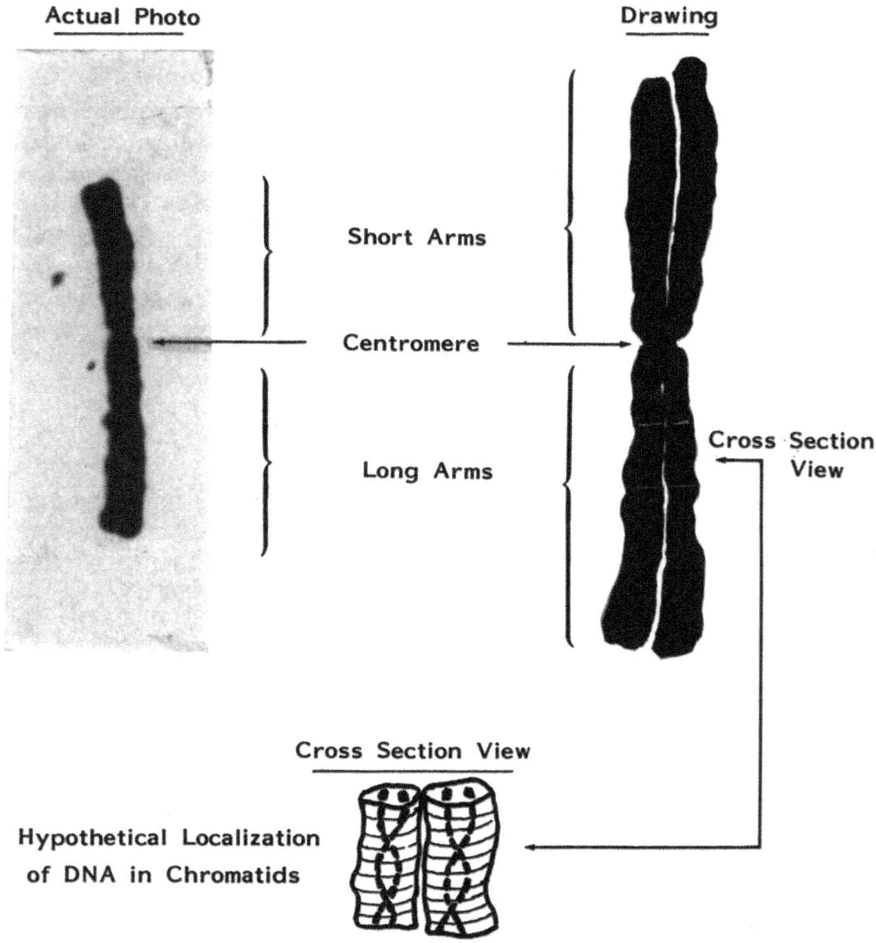


FIGURE 2.7. A condensed metaphase chromosome with chromatids. This chromosome consists of DNA and associated proteins. The presence of chromatids indicates that the chromosome is in the metaphase stage of mitosis just prior to division (see Figure 2.8). The enlarged cross section is a simplified schematic of how the DNA is packaged in a chromosome.

indicates the nucleolar organizer region of the chromosome. The nucleolar organizer region is the site of synthesis on genes for RNA used in ribosome construction. The chromosomal region beyond the secondary nucleolar constriction occasionally appears terminally disassociated from the body of the chromosome and is referred to as a satellite.[8] The location of these

| STAGE | EVENT | CELLULAR CONFIGURATION |
|---|---|---|
| Interphase | Chromatin uncondensed in nucleus. | 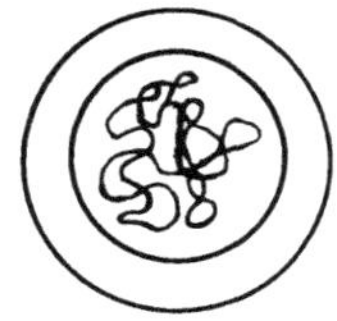 |
| Prophase | Chromatin begins to replicate and condense into visible chromosomes. Nuclear membrane disappears. |  |
| Metaphase | Replicated chromosomes visible as chromatids which arrange for cell division. |  |
| Anaphase | Chromatids are separated into two daughter cells. |  |
| Telophase | Two identical daughter cells are formed. The nuclear membrane reforms and the chromosomes uncondense and return to interphase appearance. | 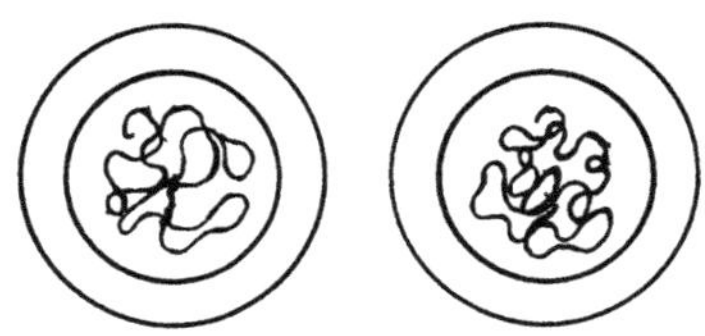 |

FIGURE 2.8. Mitosis. The stages of somatic cell division.

structures is important in chromosome identification and for the visualization of structural alterations induced by chemicals.

Other features that can be localized on the chromosomes are regions of heterochromatin (which has a low content of functional DNA) and euchromatin (which has a high content of functional DNA). The two regions are believed to represent tight and loose coiling regions of DNA, assumed to be the nonactive and active regions, respectively. Staining techniques that rely upon differential protein concentration are referred to as banding techniques and can be very useful in individual chromosome identification when the overall size and structure appear identical. Each chromosome has its own characteristic and individualized banding profiles.

The total number and array of chromosomes in a cell is referred to as the cell's karyotype. The karyotype for a species is unique and often forms the basis of cellular taxonomy (Figure 2.9). Thorough knowledge of the normal karyotype of a species is essential before entering into cytogenetic studies in which structural or numerical alterations are scored.

## The Mitotic Cell Cycle

The cell cycle is oriented toward DNA replication, followed by its division and distribution to each of the two new daughter cells. At the molecular level, DNA replication is an enzymatic process involving a series of enzymes that open the DNA helix, synthesize complementary copies by polymerization of appropriate nitrogenous bases, and seal the synthesized strands into a complete new macromolecule. This process is followed by separation of the two new helices. This phenomenon is called semiconservative replication, since one strand of each new helix is conserved and forms the template upon which the new strand is polymerized (Figure 2.10). It is extremely rapid and accurate (less than one error per million bases polymerized).

Eukaryotic cells, especially mammalian cells, have a cell cycle with distinct stages. Although the DNA in eukaryotic chromosomes is believed to replicate via the mechanism shown in Figure 2.10, organization and replication of the accessory chromosome constituents require these cells to undergo a relatively elaborate set of events identified as the "cell cycle." Knowledge of the specific activities and duration of these stages is essential in conducting and evaluating many of the tests in genetic toxicology, particularly those that measure unscheduled DNA synthesis (UDS), sister chromatid exchange (SCE), and various *in vitro* and *in vivo* chromosome abnormalities.

Table 2.4 illustrates the various stages in the cell cycle and defines the processes involving DNA/chromosomes that occur in each stage. Mitosis (M) phase is the stage at which the replicated DNA condenses into discrete chromosomes (chromosomes are not cytologically visible in other stages of the cell cycle), which are then distributed into two new daughter cells. The process of mitosis is arbitrarily divided into five discrete steps

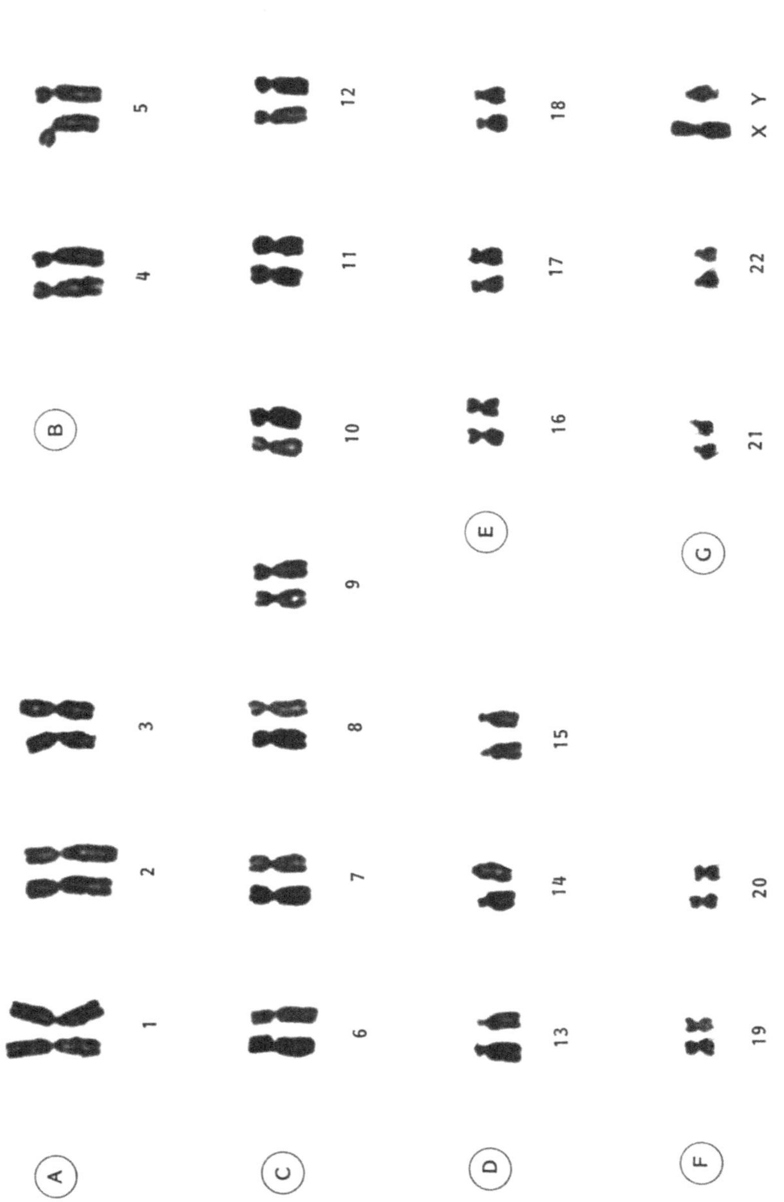


FIGURE 2.9. Human karyotype. This figure illustrates how chromosomes are grouped by size and shape to determine the karyotype of a species. Such karyotypes are used for diagnosis of certain hereditary abnormalities in humans. (Photo courtesy of H. Lebowitz.)

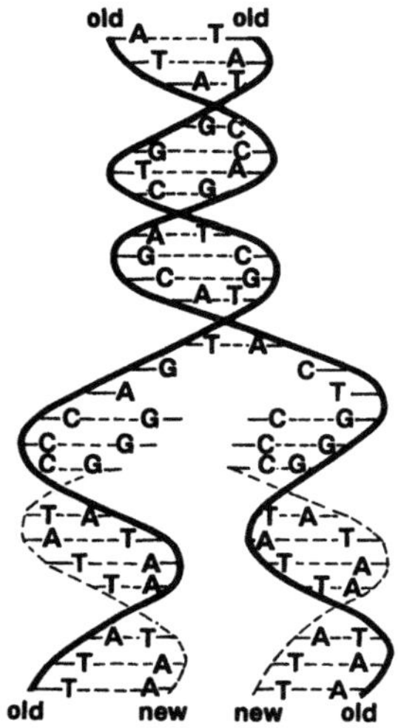


FIGURE 2.10. Structure and duplication of DNA. Semiconservative replication involves the unwinding of the DNA helix, exposing each strand. The open strands serve as templates for the enzymatic synthesis of new complementary strands of DNA. DNA polymerase α is the replicative enzyme for the process of replication. The results will be two duplicate DNA molecules. (From Levine.[9])

(Figure 2.8). Of these steps, metaphase is typically viewed in cytogenetic testing to evaluate chromosomes for breakage and rearrangements. Chemical treatment of cells with colchicine will arrest cells in metaphase for easy viewing. Anaphase has also been used for aberration detection, but the chromosome arrangements are more difficult to interpret. Very little use is made of the anaphase method in genetic toxicology.[10] Examples

TABLE 2.4
Stages of a Typical Mammalian Cell[a]

| Stage | $G_1$ | S | $G_2$ | M |
|---|---|---|---|---|
| Event | $GAP_1$ | Synthesis of DNA | $GAP_2$ | Mitosis (see Figure 2.8) |
| Relative DNA content | 1 | $1 \rightarrow 2$ | 2 | $2 \rightarrow 1$ |
| Common type of aberration observed[b] | Chromosome breaks<br>Rings<br>Translocations<br>Deletions<br>Dicentrics<br>Fragments | Mixed | Chromatid breaks<br>Quadriradials<br>Triradials<br>Fragments | |
| Relative sensitivity to genotoxic effects | Low | High | Moderate | Low |

[a] Shown as relative lengths of the total cell cycle period for *in vitro* conditions.
[b] These aberrations are the types observed when the exposed $G_1$, S, or $G_2$ cells are scored at the first metaphase (M1). If the cells are scored at M2 or M3, the aberration types are mixed and lose identity with the cell stage.

of specific chromosome changes scored in cytogenetic evaluations are described later in this chapter.

## Meiosis and Chromosome Mechanics

Meiosis occurs in germinal tissue of diploid species. The end result of this process is the reduction of the diploid ($2N$) chromosome number to a haploid ($N$) number, ensuring that each gamete has one copy of each chromosome (and gene) pair. Figure 2.11 illustrates the meiotic process in mammals. This process also includes the mechanisms of independent assortment and distribution of maternal and paternal chromosomes to the progeny, as well as chromosome recombination (crossing over). See Strickberger[11] for a more detailed discussion of these events.

Independent assortment of chromosomes and genetic recombination are two parameters of meiosis important in the randomization of genes before they are packaged as gametes (sperm or ova). Genetic recombination normally occurs during meiosis and involves reciprocal exchanges between chromosomes of a homologous pair adding to the randomization of genes on chromosomes during gametogenesis. Without these processes, genetic variability would be limited to rare mutational events that survive natural selection. While independent assortment and recombination are critical to the understanding of the distribution of genes to gametes and inheritance patterns of genetic traits, they are not essential to most aspects of genetic toxicology testing.

In genetic toxicology, knowledge of the process of meiosis is important when conducting *in vivo* assays such as those employed to detect dominant lethality, specific-locus mutation, and heritable translocations. It is essential to know the length of time required for a chromosome set to pass from the initial stage of meiosis (gonia cells) to the formation of mature sperm or ova, since acute exposure to a mutagenic substance may produce a stage-specific effect. For example, if a test substance is genetically active only on chromosomes undergoing the meiotic metaphase I cell division, an effect would not be evident unless the exposed male organism is mated to females over a period of time sufficient to permit the damaged cells to traverse through meiosis I and II cell divisions to mature sperm (Figure 2.12). In a typical rodent dominant lethal study using male mice or rats, the results of the stage-specific response described above would not be seen until the 5th or 6th week of mating. Conversely, if the test sample is active only in mature spermatozoa, the response in a dominant lethal test would be evident within the first 2 weeks of mating. Little or no effect would be seen after the 3rd or 4th week, since the sperm would have been premeiotic cells at the time of chemical exposure and refractile to its genotoxic properties. Similar considerations

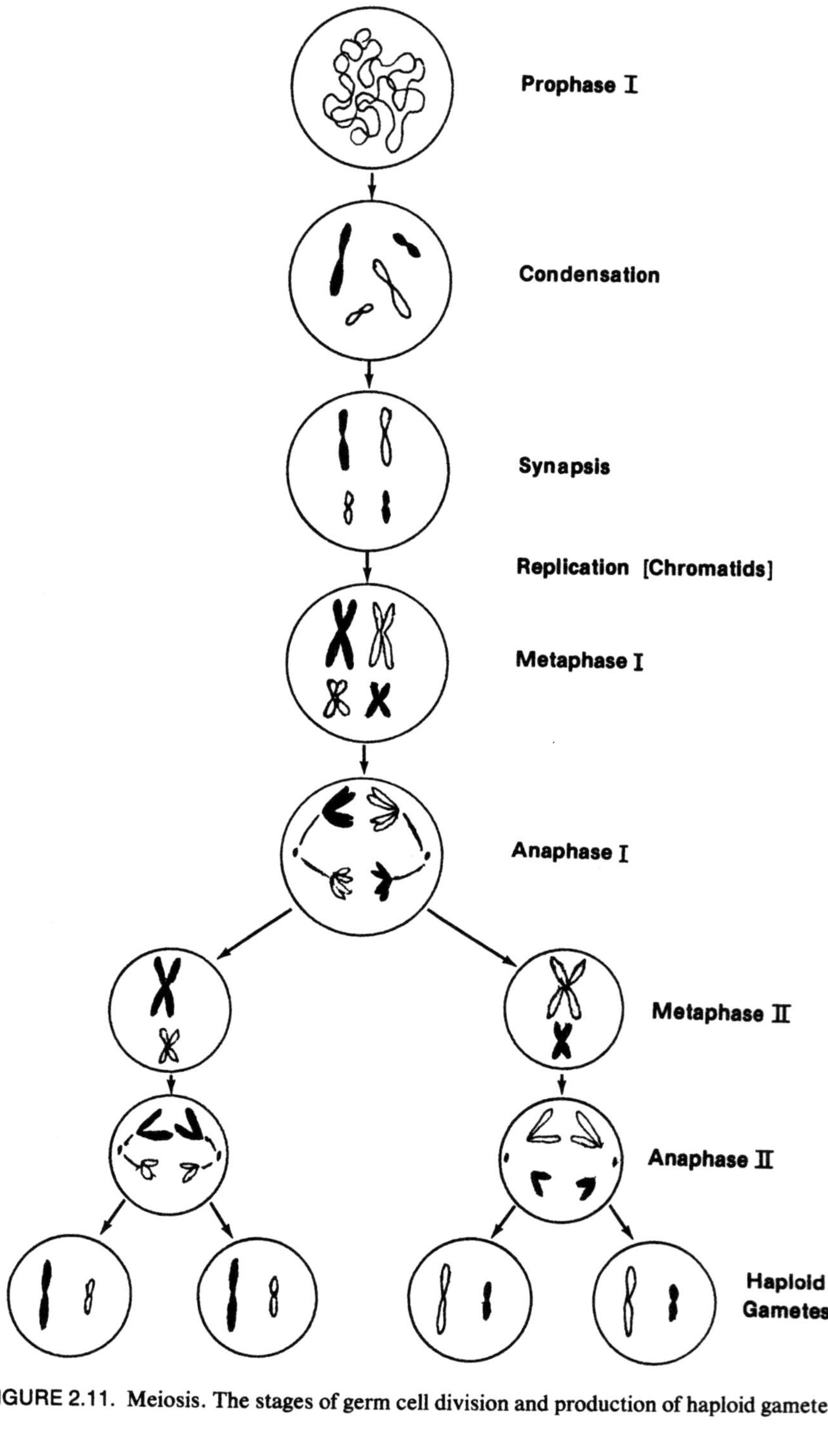


FIGURE 2.11. Meiosis. The stages of germ cell division and production of haploid gametes.

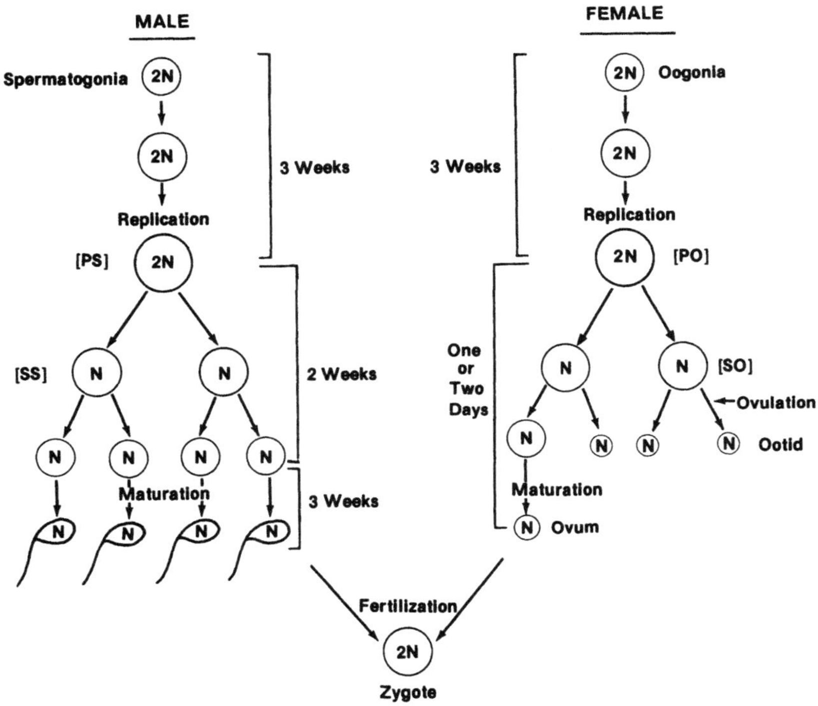


FIGURE 2.12. Passage of gametes through meiosis in rodent species. Male and female stages differ in the number of gametes produced and in temporal sequence. Sperm are produced and cleared continuously. The female germ cell remains as a primary oocyte until just before ovulation. Thus, female gametes may experience multiple exposures *in situ* before release. PS/PO, Primary spermatocyte or oocyte; SS/SO, secondary spermatocyte or oocyte.

operate for test methods with *Drosophila*, but generally a period of 8–10 days is sufficient to cover the entire meiotic process in this test species.[12]

Alternative dosing methods have been proposed for dominant lethal and mouse heritable translocation assays to reduce the number of animals required in the study by eliminating the sequential mating procedures routinely employed and allowing all exposed cells to accumulate. The proposed alternatives do not, however, identify stage-specific responses. Practical considerations were responsible for the modifications, since the cell stage on which a mutagen acts becomes academic to the toxicologist concerned primarily with detecting an effect. Under the alternative approach, males are dosed over the entire meiotic process, followed by two mating sequences. The rationale is that the chromosomes that will form mature gametes have been continuously exposed during all stages of their cycle through the meiotic process, and thus any stage-specific alterations

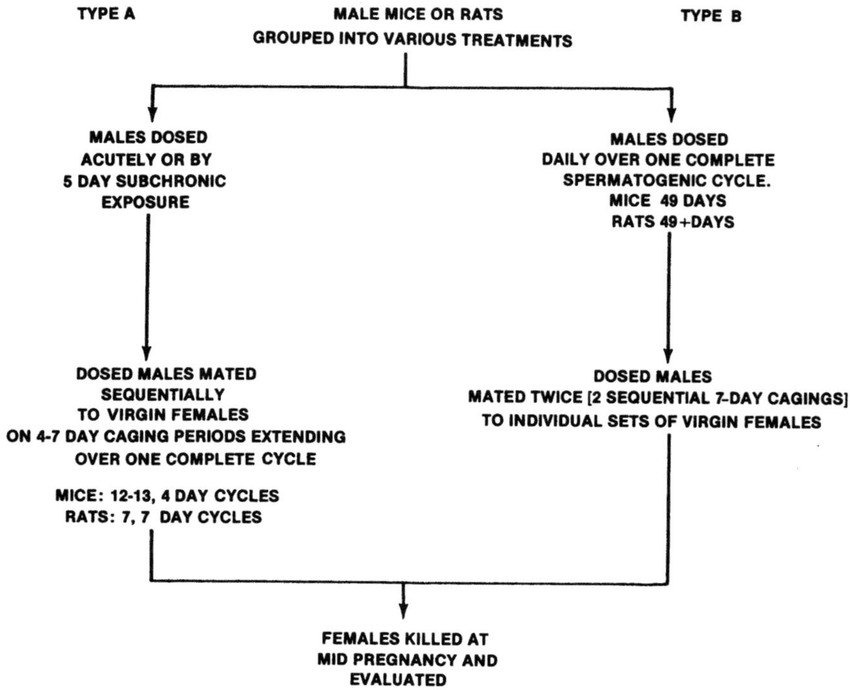


FIGURE 2.13. The male dominant lethal assay. Alternate protocols using continuous dosing versus acute exposure. (From Brusick.[13])

will be present and expressed in the matings following continuous dosing. The continuous dosing method is the only practical approach for mouse heritable translocation studies, since sequential weekly matings in this assay would involve prohibitively large numbers of animals. Exposure longer than one complete meiotic cycle is not advantageous since the same pattern of responses would be repeated. The two methods of dosing are compared in Figure 2.13.

Testing that involves exposure of females during the meiotic process is complicated by the fact that all ova in the female mammal proceed through meiosis division I and are arrested before meiosis division II. The maturing ova go through division II just before maturation and are released. A few dominant lethal studies have been conducted in female rodents, but the technical difficulties severely limit this type of testing.[14,15] Additional reasons for not using female animals in dominant lethal studies are that (1) oocytes are generally less sensitive to chemicals than are sperm and (2) one cannot easily discriminate between cytotoxic effects impairing ovulation and true genetic effects that result in preimplantation loss. Artifactual preimplantation loss also occurs in male-treated dominant lethal tests, but would be expected to be a more reliable indication of genotoxicity than that found in the female-treatment procedure. Good female dominant lethal studies can be performed by taking the time to identify the estrous cycle in the females to be treated and then setting up a regimen in which the treatment can be commenced following ovulation, so that one will be treating cells more sensitive than the oocytes.

## DNA ALTERATIONS RESULTING IN GENOTOXIC EFFECTS IN CELLS: MECHANISMS AND CATEGORIZATION

DNA replication and DNA repair are not perfect processes and, in rare instances, alterations in the genetic integrity of a cell do occur spontaneously. The occurrence of the changes follows a predictable rate per gene and forms the basis of "background" mutation frequencies. Spontaneous DNA changes occur at both the nucleotide and chromosome levels. Most mutations are detrimental to cells and are rapidly lost from the cell population; under the appropriate environmental selection pressures, however, some mutations may be neutral (result in neither beneficial nor deleterious traits) or even beneficial.

The genetic composition of human beings has evolved over several million years and exists in a delicate balance. Introduction of random changes, even minor ones, into this carefully evolved system can result in serious diseases, malformations, or death. Although evolution depends on mutation for the introduction of variability into populations, too many changes in too short a time period could seriously reduce the viability of the human species. The toxicologic consequences of mutation in humans will be discussed in Chapter 3.

### A Classification Scheme for Genotoxic Effects

DNA damage consists of two broad categories: visible effects detectable through cytologic analysis of chromosomes (macrolesions), and nonvisible changes, which occur at the nucleotide level (microlesions). The specific types of DNA damage falling into these two categories are shown in Figure 2.14.

Microlesions consist of base-pair substitutions or base-pair addition/deletion changes. Base-pair substitution mutations result from qualitative changes in the nucleotide composition of a codon. For example, if in the RNA codon 5′ GAG 3′, which codes for glutamic acid, the initial guanine is substituted by adenine to form 5′ AAG 3′, the RNA codon will then specify lysine (see Figure 2.2). If the substitution of lysine for glutamic

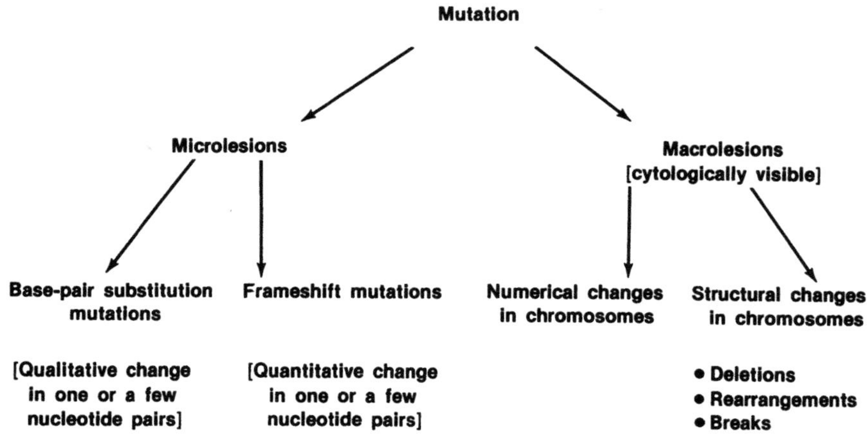


FIGURE 2.14. Classification of the molecular changes in DNA that occur as a result of mutation. (From Brusick.[13])

acid in the polypeptide gene product is incompatible with normal function, the result in a loss in the activity of that gene (probably mutation). The most common mechanism for microlesion production is the mispairing of bases as a consequence of chemically induced damage.

The expected configurations of base pairs are shown in Figure 2.15. Adenine/thymidine form two hydrogen bonds and guanine/cytosine form three. Hydrogen bonds are weak electrostatic forces involving oxygen and nitrogen atoms at specific sites on the purine and pyrimidine molecules. If electrophilic chemical species covalently bind to portions of the DNA bases involved in the formation of hydrogen bonds (Figure 2.16A), these covalently bond species (adducts) might result in mispaired bases and eventually substitution of one base pair for another in the DNA mol-

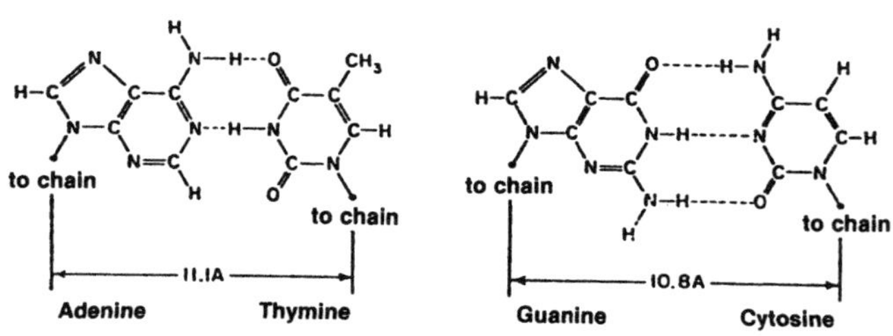


FIGURE 2.15. Normal base pairing for DNA bases. Adducts that attach at N or O atoms involved in hydrogen bonding (– –) may produce abnormal pairing of adenine with cytosine or guanine with thymidine. These abnormal pairs may then produce mutation (see Figure 2.16A).

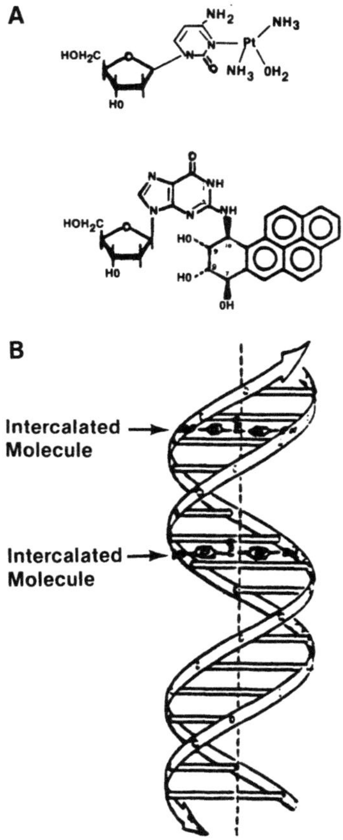


FIGURE 2.16. (A) Covalent binding (adduct-forming mutagens). Two examples of adduct formation, involving (above) *cis*-platinum adduct of cytosine at the NH position involved in H bonding with guanine, and (below) a benzpyrene metabolite adducted to guanine at the NH position involved in H bonding with cytosine. (B) Intercalation (non-adduct mutagens): the hypothetical mechanism by which simple intercalaters attach to DNA via weak electrostatic forces. The intercalated molecules may interfere with either DNA repair enzymes or replication enzymes, resulting in base-pair additions or deletions, which in turn produce frameshift mutations.

ecule (Figure 2.17). Substitutions of purine for purine and pyrimidine for pyrimidine are designated transitions, while substitutions of purine for pyrimidine or pyrimidine for purine are designated transversions.[16] Transitions as well as transversions have been found in mutant DNA sequences isolated from test organisms.

Base-pair addition/deletion mutations, also called frameshift mutations, result from the addition or deletion of one or a few nucleotide pairs from the nucleotide complement in a gene (Figure 2.16B).[17] Since the codon sequence of a gene is nonpunctuated, the loss or gain of a single base pair changes the reading frame of the gene; hence, one has a frameshift mutation. This type of change is illustrated in Figure 2.18.

Base-pair substitutions and frameshift mutations are probably induced by distinctly different mechanisms and generally by distinctly different classes of chemical mutagens. Both types of changes are considered important toxicologically, and like stage-specificity in the dominant lethal

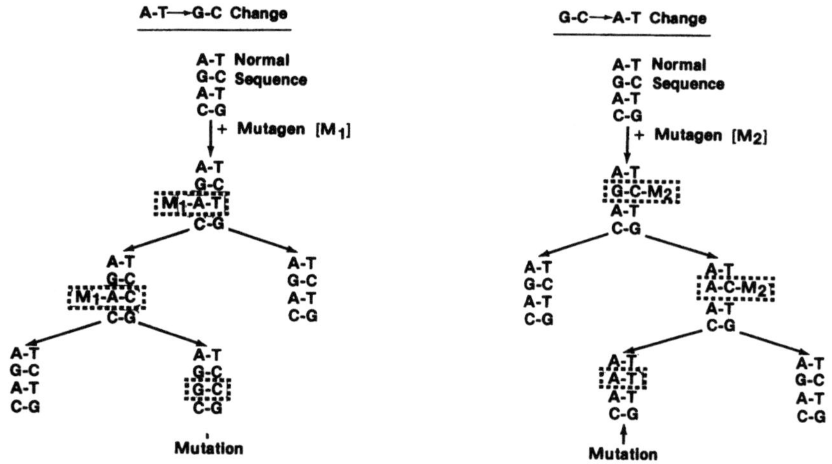


FIGURE 2.17. Mechanisms for base-pair substitution induction. Changes in base sequence brought about by two different mutagens, M1 and M2, are shown. The mutagens differ in their affinities for reaction site. One reacts with purines (A-M1 reaction) and the other with pyrimidines (C-M2 reaction). The proposed mechanisms produce one mutant and one normal cell following induction, in what is termed mosaicism.

assay, their distinction experimentally is of greater academic than toxicologic interest.

Macrolesions can be subdivided into changes in chromosome number (gain or loss of single chromosomes or sets of chromosomes) and changes in chromosome structure (breaks, deletions, rearrangements). Each specific type of chromosome change has a characteristic designation, so that a reasonably high degree of uniformity can be maintained in scoring for these changes. Most cytogenetic studies consist in evaluating mitotic metaphase chromosomes. A glossary of terms and descriptions used in metaphase analysis is provided in Table 2.5. Figure 2.19 illustrates many of the aberrations described.

Variations in chromosome number can result from incomplete dissociation of single or entire sets of chromosomes at metaphase. This may result in cell aneuploidy, monosomy, trisomy, etc., causing serious effects in intact organisms. Some relatively common human disorders, such as Down's syndrome, result from variations in chromosome number.

Classically, chromosome aberrations involve structural changes classified according to their time of formation relative to the cell cycle (Table 2.4). Chromatid breaks exhibit one normal and one broken chromatid of a pair, whereas in chromosome breaks both chromatids are broken at identical locations. The same type of relationship is involved in chromatid

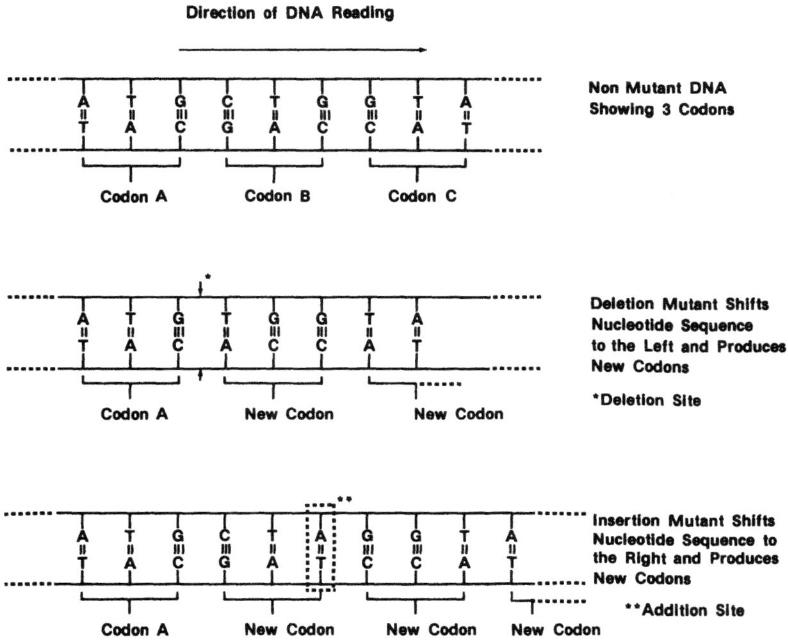


FIGURE 2.18. Mechanisms for frameshift mutation induction. The production of altered codon sequences by insertion or deletion of base pairs will lead to changes in amino acid sequences in the gene product and probably result in defective gene activity.

and chromosome rearrangements. The production of either chromatid and/or chromosome aberrations by an agent depends on the nature of the clastogen (chromosome-breaking agent) and the cell cycle stage the target cell was in at the time of exposure. The majority of chromosome-type effects are described from lesions induced in $G_1$. A $G_2$ exposure generally results in chromatid aberrations. Most chemical mutagens, however, induce chromatid-type aberrations independent of the cell-cycle stage, provided the cells are examined in the first mitosis after treatment (M1). In cytogenetic analyses, exposed cells can be collected at several time intervals following an acute exposure. Protocols should be structured to ensure that the number of cell cycles that have occurred posttreatment is determined and correlated to the scoring process. The greatest frequency of structural aberration should be expressed in M1 cells. Preliminary screening of treated cells for bromodeoxyuridine (BrdU) incorporation (SCE staining method) can define the best time for cell harvest.

An issue of considerable controversy in cytogenetic evaluations is the significance or lack of significance of gaps in chromatids and/or chro-

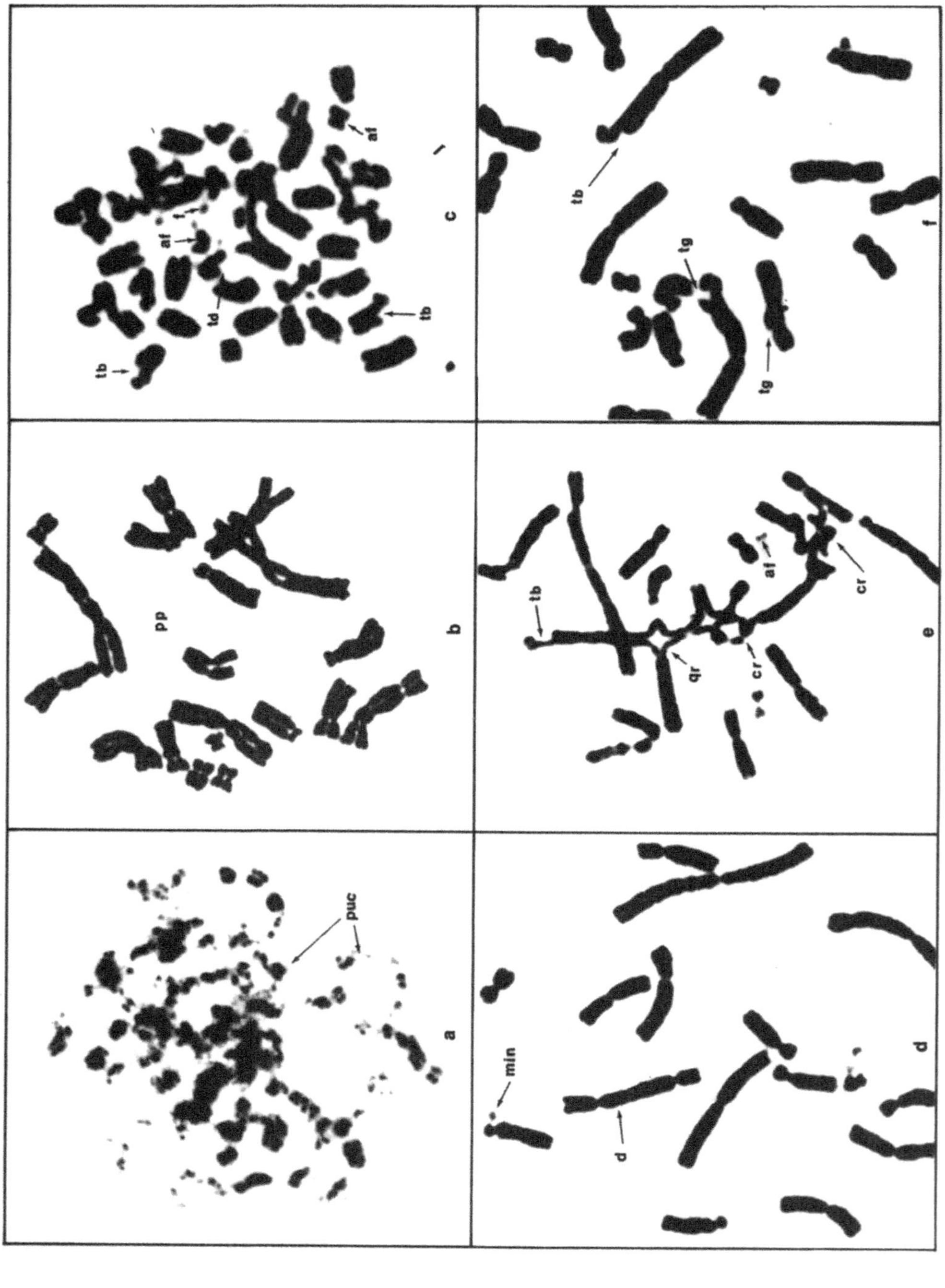
a
puc
b
pp
c
tb
af
f
td
d
min
e
qr
cr
tg

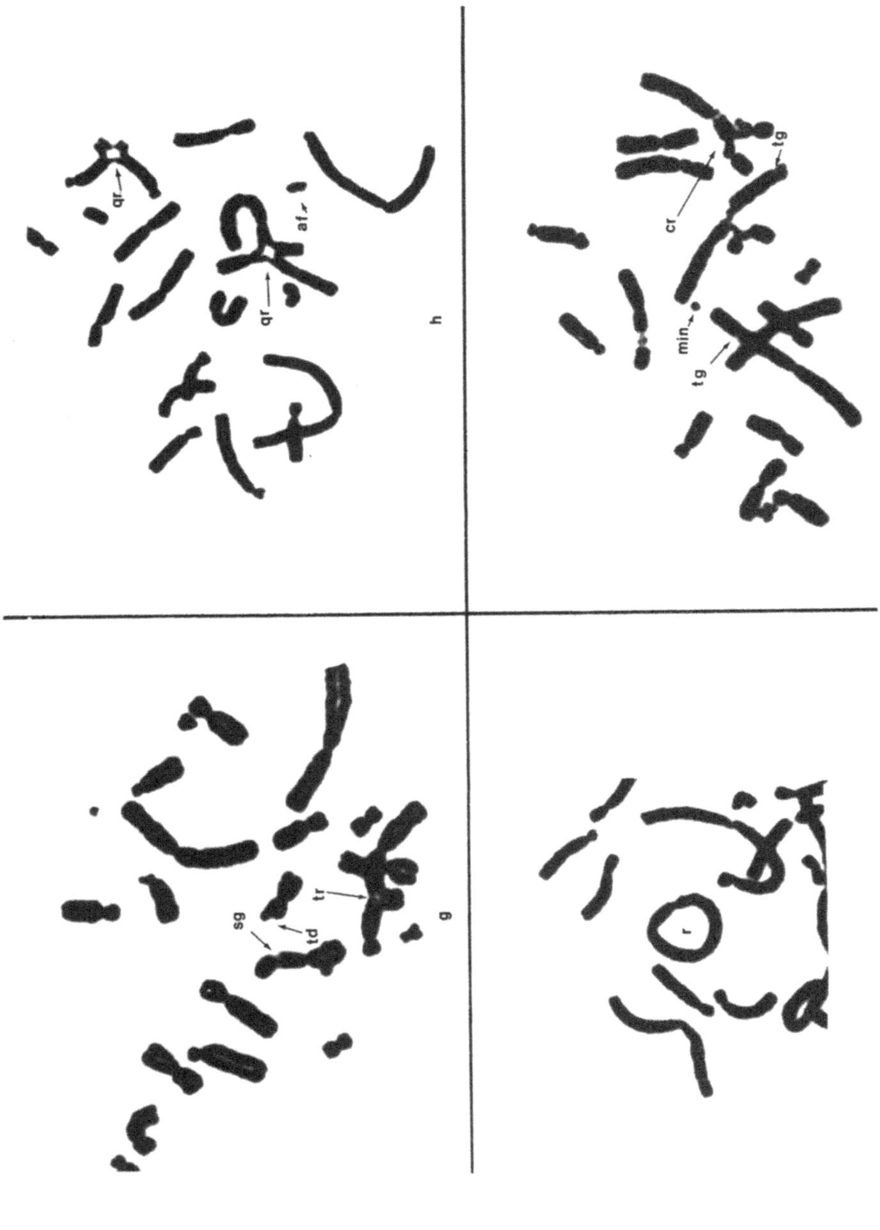


FIGURE 2.19. Examples of typical chromosome aberrations scored in mammalian somatic cells. The descriptions and alteration symbols are given with reference to a metaphase spread where the alterations can be visualized.

TABLE 2.5
Definitions of Aberrations

| Term | Definition | Example[a] |
|---|---|---|
| Chromatid gap (tg) | An achromatic region in one chromatid, the size of which is equal to or smaller than the width of the chromatid | f, j |
| Chromatid break (tb) | An achromatic region in one chromatid larger than the width of the chromatid; it may be either aligned or unaligned | c, e, f |
| Chromosome gap (sg) | Same as tg, only in both chromatids | a, f |
| Chromosome break (sb) | Same as tb, only in both chromatids | Not shown |
| Chromatid deletion (td) | Deleted material at the end of one chromatid | c, g |
| Fragment (f) | A single chromatid without an evident centromere | c |
| Acentric fragment (af) | Two aligned (parallel) chromatids without an evident centromere | c, e, h |
| Translocation (t) | Obvious transfer of material between two or more chromosomes | Not shown |
| Triradial (tr) | An abnormal arrangement of paired chromatids resulting in a triarmed configuration | g |
| Quadriradial (qr) | An abnormal arrangement of paired chromatids resulting in a four-armed configuration | e, h |
| Pulverized chromosome (pu+) | A spread containing one fragmented or pulverized chromosome | Not shown |
| Pulverized chromosomes (pu+) | A spread containing two or more fragmented or pulverized chromosomes, but with some intact chromosomes still remaining | Not shown |
| Pulverized cell (puc) | A cell in which all the chromosomes are totally fragmented | a |
| Complex rearrangement (cr) | An abnormal translocation figure that involves many chromosomes and is the result of several breaks and mispaired chromatids | c, j |
| Ring (r) | A chromosome that is a result of telomeric deletions at both ends of the chromosome and the subsequent joining of the ends of the two chromosome arms | i |
| Minute (min) | A small chromosome that contains a centromere and does not belong in the karyotype | d, j |
| Polyploid (pp) [endoreduplication] | A cell in which the chromosome number is an even multiple of the haploid number $N$ and is greater than $2N$ | b |
| Hyperdiploid (h) | A cell in which the chromosome number is greater than $2N + 1$, but is not an even multiple of $N$ | Not shown |
| Dicentric (d) | A chromosome containing two centromeres | d |

[a] Examples illustrated in Figure 2.20.

mosomes. Some investigators feel that identification of gaps is extremely subjective and that they do not constitute true aberrations.[18] This group of investigators has suggested that gaps may be quantified and reported, but should not be used in the evaluation of the study. Other investigators routinely include gaps among their aberration totals and use this information in evaluation of the test results. The EPA guideline protocols for cytogenetic studies are consistent with the former position on this subject and probably represent a consensus of the cytogenetic community; until this issue is resolved, however, conflicting results may be reported for cytogenetic evaluations.

Cytogenetic evaluations cannot be used alone to define mutagens since not all mutagens produce DNA lesions that lead to the formation of chromosome aberrations. Mutagens that act via certain types of base substitution reactions (base analogs, hydroxylamine, and some monofunctional alkylating agents) or those that are pure DNA-intercalating frameshift mutagens (acridine and anthraquinone dyes) are not effective clastogens. Bifunctional or polyfunctional alkylating agents, which are good inter- and intrastrand cross-linking agents [nitrogen mustard and triethylenemelamine (TEM)], are the most effective type of clastogenic agents known.

A cytogenetic technique currently in wide use is the analysis for SCEs. This phenomenon was originally described by Taylor in 1957, but analysis for SCEs on a routine basis only became possible following the development of simple staining techniques that differentiate sister chromatids.[19,20] The mechanisms proposed to explain the staining technique and formation of SCEs are outlined in Figure 2.20. SCE analysis appears to be a very good screening test and an excellent dosimeter for exposure, but the biological significance of SCEs as a genotoxic lesion is not known at the present time.

Another screening technique developed to assess the induction of chromosome damage is the test for production of micronuclei. Schmid[21] and Heddle[22] developed this test independently, the distinguishing feature being a minor difference in the preparation of the slides. The theoretical basis for micronuclei rests with the hypothesis that broken chromosomes or chromatid fragments may lag behind intact chromosomes during the anaphase step of mitosis. During telophase daughter nuclei are formed. If the broken and lagging chromatin is not included in the main nucleus during telophase, micronuclei are formed in the cytoplasm (Figure 2.21). Thus, it is believed that the frequency of cells containing micronuclei following chemical treatment is indicative of clastogenic activity.[23] While micronuclei may be the results of broken chromosomes or chromatids that produce lagging anaphase fragments, it is possible that chromatid damage may not be expressed after a single cell cycle (e.g., chromatid

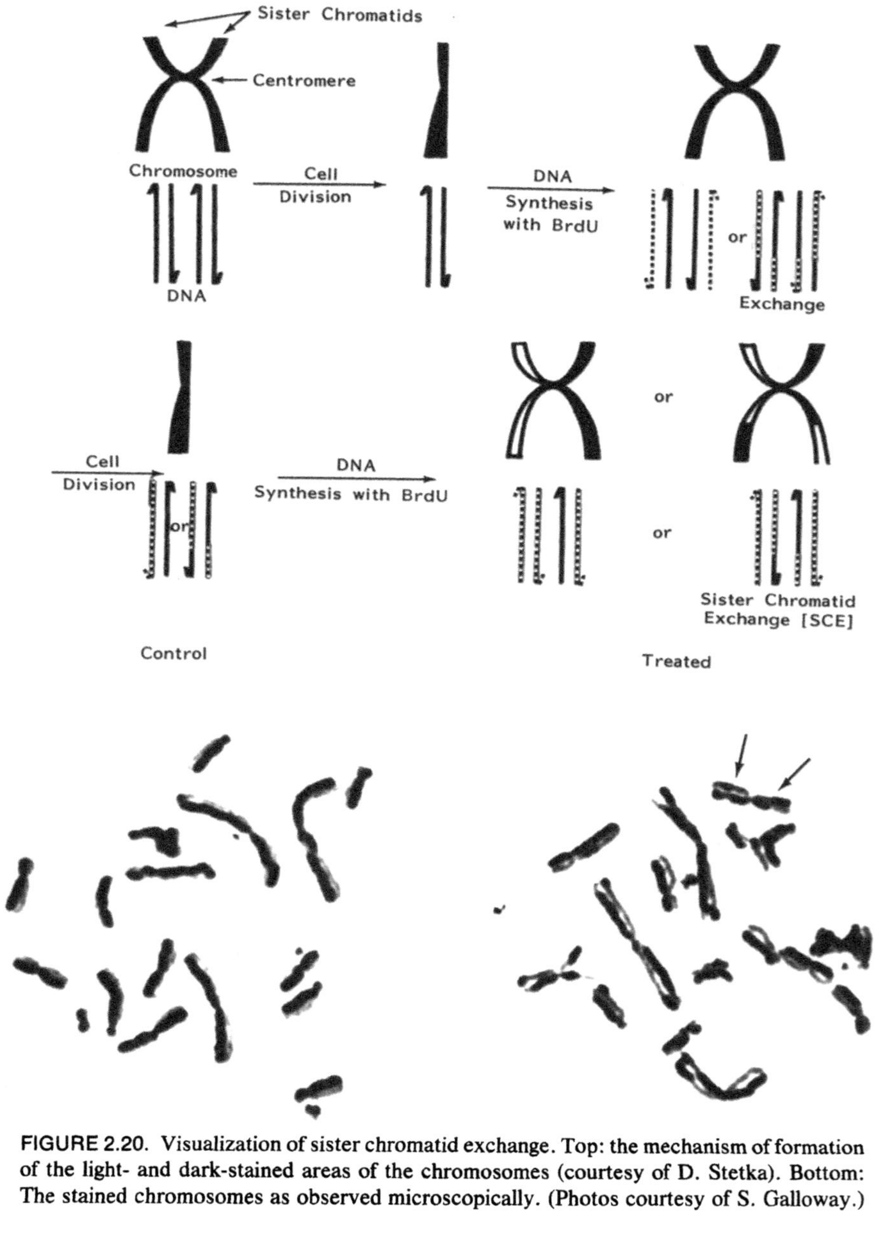


FIGURE 2.20. Visualization of sister chromatid exchange. Top: the mechanism of formation of the light- and dark-stained areas of the chromosomes (courtesy of D. Stetka). Bottom: The stained chromosomes as observed microscopically. (Photos courtesy of S. Galloway.)

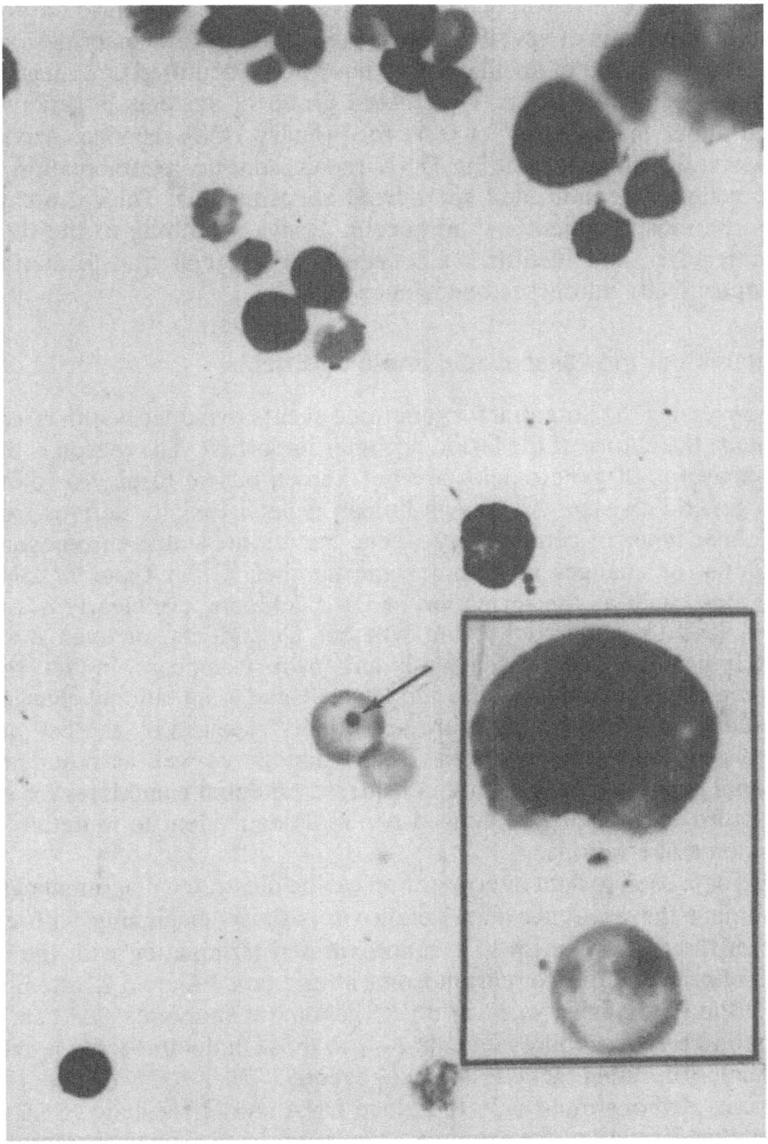

FIGURE 2.21. Micronuclei observed in mouse polychromatic erythrocytes. An enlarged view is also shown in inset. The micronucleus is shown as a darkly stained body in the cell. (Photo courtesy of H. Lebowitz.)

breaks) and induced micronuclei may be missed using protocols with a single kill time. The study designs recommended for use at the present time are considerably different from that Schmid proposed in 1973.

Besides the tests that specifically measure microlesions (also referred to as point mutations or specific-locus gene mutations) and macrolesions, a third, diverse group of ancillary tests have been identified that measure other types of genotoxicity. This mixed group of systems is generally grouped under the category of tests for primary DNA damage. Among these tests are those measuring DNA repair, mitotic recombination or mitotic gene conversion, and spermhead abnormalities. Table 2.6 categorizes the most frequently used genetic assays according to the three basic categories and identifies a fourth category, cell transformation, which specifically addresses oncogenicity.

## Significance of the Classes of Genotoxic Effects

In assessing the potential for genotoxic agents to induce health effects in humans, the nature of the lesion becomes important. The reason is that many measures of genotoxicity are not known *per se* to play a role in human genetic disease. All known human genetic toxicity derives from one of three types of genotoxicity—gene mutations, stable chromosome aberrations, or changes in chromosome number. Other types of DNA interactions, such as the formation of DNA adducts, are clearly events that can lead to genotoxicity, but whether all adducts, or even most, ultimately produce genetic damage is unknown. It appears, in fact, that for a given chemical only specific adducts will lead to mutational changes; other adducts formed by exposure are rapidly repaired by the cell and have no genotoxic consequences. DNA adducts, as well as other premutational genotoxic lesions, are considered potential candidates for genetic hazard, but cannot be viewed *per se* as equivalent to mutation or chromosome aberrations.

This approach to data interpretation can be illustrated diagramatically by assuming the existence of a genotoxic pathway beginning with exposure of the test organism to a genotoxin and terminating with the induction of gene mutation or chromosome alterations. Figure 2.22 attempts to show the relationship of a variety of genotoxic endpoints that can be measured in test procedures (events 1–4) to those endpoints known to be responsible for human genetic disease (events 5–7).

Agents demonstrated only to induce DNA strand breakage, SCE or DNA adduct formation, for example, should not be viewed with the same genetic hazard potential as agents established as gene or chromosomal mutagens until the definitive tests have been conducted. In other words, a genotoxic agent must be shown to induce those lesions known to be

TABLE 2.6
Basic Categories of Genetic Assays and Some Examples[a]

| Gene mutation | Chromosome aberration | Primary DNA damage | Oncogenic transformation |
|---|---|---|---|
| Bacteria forward and reverse mutation<br>Ames/*Salmonella*<br>*Escherichia coli*: $WP_2$ uvrA−<br>*Escherichia coli*: K12 forward mutation<br>Yeast forward and reverse mutation<br>*In vitro* mammalian cell forward mutation<br>Mouse lymphoma TK assay<br>CHO HGPRT assay<br>V79 HGPRT assay<br>*Drosophila*: SLRL<br>Mouse: somatic mutation test (spot test)<br>Specific locus assay in mice | *In vitro* cytogenetics<br>Mouse bone marrow<br>Mouse micronucleus<br>*Drosophila*: X and Y loss<br>*Drosophila*: dominant lethal<br>*Drosophila*: heritable translocation<br>Mouse: dominant lethal<br>Rat: dominant lethal<br>Mouse: heritable translocation | *Escherichia coli*: polA+/polA− assay<br>*Bacillus subtilis*: rec+/rec− assay<br>Yeast $D_3$, $D_4$, $D_5$, or $D_7$ somatic recombination<br>UDS<br>WI-38 cells<br>Rat primary hepatocytes<br>SCE analysis | *In vitro* cell transformation:<br>BALB/c 3T3 cells<br>SHE cells<br>Mouse 10T½ cells |

[a] SHE, Syrian hamster embryo. TK, Thymidine kinase. CHO, Chinese hamster ovary. HGPRT, Hypoxanthine-guanine phosphoribosyl transferase gene. SLRL, Sex-linked recessive lethal.

responsible for human genetic disease before it can be considered hazardous. A chemical that induces gene or chromosome mutations in the gametes of test mammals should carry sufficient evidence for adverse findings to warrant regulatory action in a fashion similar to that taken following positive findings for cancer or terata in respective animal models.

This approach to test analysis, however, is not relevant to genetic correlative diseases such as tumor induction. At the present time there is no reason to select one type of genotoxic endpoint over others as the

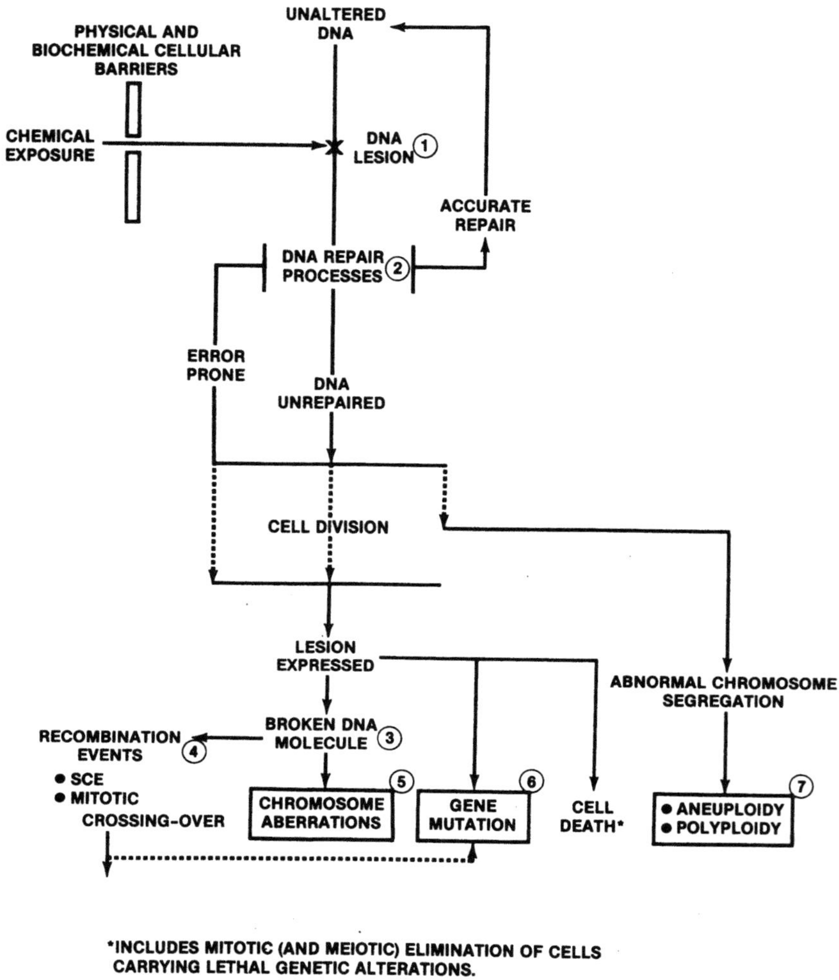


FIGURE 2.22. The hypothetical relationship between the internal interaction between a biologically active molecule and DNA damage that may ultimately lead to a genetic endpoint responsible for human disease or dysfunction. The three endpoints known to produce human genetic disease are identified by the closed boxes. Steps along the pathway identified by the numbers 1–4 represent general classes of genotoxic endpoints measured in commonly used genetic toxicology assays.

most predictive test for carcinogenesis *in vivo*. All test methods for genotoxicity that show good empirical correlations with cancer would therefore be considered relevant, regardless of the nature of the genotoxic lesion they detect.

## REPAIR OF DNA DAMAGE

To maintain the fidelity and integrity of genetic information, several types of enzymatic DNA repair processes developed during evolution. DNA is the only molecule in living organisms with a capacity for self-repair. The common feature of repair processes is the ability to remove and replace damaged segments of DNA. Therefore, if a DNA lesion induced by a mutagen can be repaired prior to fixation or stabilization, the net effect of the DNA insult may be nil. This is especially true following low-level exposure where excision repair enzymes are not fully saturated by excessive numbers of damaged DNA sites. Systems measuring some parameter of DNA repair[24–26] have been used as direct measurement of primary DNA damage. Since all normal organisms are capable of some type of repair following chemical insult, a stimulation in the level of repair activity following chemical treatment at sublethal concentrations is a good general indicator that the test sample has DNA-directed toxicity.

Four types of repair mechanisms have been identified. The general characteristics of each are summarized in Table 2.7. Excision repair and, to a certain extent, postreplication repair appear to be the two processes involved with elimination of most chemically induced DNA lesions, and both contribute greatly to the net genetic effect following exposure of organisms to various genotoxic agents.

Excision repair operates prior to replication of the DNA, whereas postreplication repair allows replication to proceed around lesions that might otherwise be lethal. These two types of repair are also believed to be inducible as well as constitutive.[27] Both processes result in error-free and error-prone repair (Table 2.8). Inherent variation in repair capacity between organisms can lead to differential susceptibility to genotoxic effects. Hart and Setlow reported on the correlation between DNA excision repair capability and life span in several mammalian species[28]: there appears to be increased repair in species with longer life spans. This relationship could well play an important role in the extrapolation of genetic effects, since most risk assessment models employ a conservative approach and assume a similar response for both rodents and humans.

In the context of genetic toxicology, repair processes can play potentially critical roles at low exposure levels. Evidence from prokaryotic systems indicates that once premutational lesions have been induced in

TABLE 2.7
A Summary of Four DNA Repair Mechanisms

| Excision repair | Postreplication repair | Photoreactivation | SOS repair |
|---|---|---|---|
| Damaged DNA areas excised and new sequences resynthesized in gaps | Only occurs during DNA replication | Specific for the repair of ultraviolet light-induced damage (pyrimidine dimerization) | Appears to be confined to prokaryotic organisms such as bacteria |
| Multiple enzymatic functions involved in repair, including recognition of damaged sites, nuclease base removal, polymerase base replacement, and ligase reattachment of DNA ends | Enzymatic resynthesis is involved | Enzymatic action in the presence of visible light opens the dimers to monomers | The repair system is not normally present, but is induced by DNA damage that cannot be repaired by the normal excision or postreplication repair processes |
| Involved in human DNA repair | Involved in human DNA repair activity | Not easily demonstrated in human cells | Recent studies have established an SOS-like repair in some mammalian cells |
| Broad range of DNA lesions repaired, including altered bases, larger lesions, or cross-linking of DNA | Repair involves lesions which result in replication gaps such as altered bases or larger single-stranded lesions | Little or no relationship to repair of chemical damage | |
| | May involve recombinational events | | |

TABLE 2.8
Definitions of Properties of Excision and Postreplication Repair Processes

| Property | Description |
|---|---|
| Constitutive | The level of repair enzymes is fixed at a certain level of production; increased incidence of DNA lesions does not lead to greater repair enzyme activity |
| Inducible | Repair enzyme levels respond to the number of DNA lesions; for example, chronic pretreatment of cells with low levels of alkylating agents enhances the repair of lesions with a higher "challenge" level of the same agent |
| Error-free | The DNA repair process consists of an accurate base replacement, resulting in no net change in the DNA once the repair process is complete |
| Error-prone | The repair process is not accurate and aberrant DNA sequences are produced by the repair polymerase; new mutations may result as a consequence |

the DNA, both error-prone and error-free repair pathways are active, the results depending upon the degree to which different pathways are used. The factors that determine whether error-prone or error-free pathways will predominate are the species, cell type, chemical mutagen, and specific DNA lesion induced.[29] Some data suggest that the error-free repair pathways predominate at low exposure levels and the error-prone pathways only come into play following saturation of the error-free enzymes. Thus, repair processes can produce rather complex dose–response curves, depending upon many variables. Methods to incorporate these data into risk or threshold analyses are presently not available.

Because of the influence that genetic background has on repair capacity, serious consideration must be given to determining the quantitative impact that DNA repair systems have in the assessment and extrapolation of genotoxic risk.[28,30] An informative review of the evidence for thresholds from genotoxic agents[31] addresses the potential for repair to modify dose–response curves at the low-dose end. Theoretical assumptions and data from studies of repair mechanisms tend to support the belief that at low exposure levels error-free removal of alkyl groups from DNA can be virtually 100% effected. Thus, one observes survival shoulders and nonlinear kinetics for mutation induction in repair-proficient cells and the loss of apparent "no-effect" regions in repair-deficient cells.

The authors of the review on thresholds assessed the dose–response curves available across prokaryotic and eukaryotic species and placed the responses into the following categories:

1. Threshold suggested. One or more increasing nonzero doses show no mutagenic effect, but are followed by clear response at higher dose, or the intercept of a dose–response curve falls well below zero on the response axis.
2. Threshold possible. The intercept is negative, but uncertain.
3. Threshold absent. The response extrapolates to zero on both coordinates or has a positive intercept on the response axis.

Data from the entire set of analyses was interpreted as showing no threshold in 77% of the studies, a possible threshold in 13%, and a suggested threshold in 10%. Of the chemicals reviewed, 83% were alkylating agents. The authors concluded that while many factors, including repair, could greatly influence mutation induction kinetics, especially at low exposure levels, the data are too complex to allow generalizations about their presence or absence. Each case must be evaluated independently and adequate assessment will require good dose–response data. Therefore, the initiation of DNA lesions of all classes must be considered as having some potential to lead to heritable damage or other correlative DNA toxicity.

## REFERENCES

1. Benzer, S.: Fine structure of a genetic region in bacteriophage. In *Papers on Genetics: A Book of Readings* (L. Levine, ed.), C. V. Mosby, St. Louis, pp. 287–294, 1971.
2. Starlinger, P.: DNA rearrangements in procaryotes. *Annu. Rev. Genet.* **11**:103–26, 1977.
3. Yanofsky, C., Drapeau, G. R., Guest, J. R., and Carlton, B. C.: The complete amino acid sequence of the tryptophan synthetase A protein (α subunit) and its colinear relationship with the genetic map of the A gene. In *Papers on Genetics: A Book of Readings* (L. Levine, ed.), C. V. Mosby, St. Louis, pp. 335–337, 1971.
4. Kleinhofs, A., and Behki, R.: Prospects for plant genome modification by nonconventional methods. *Ann. Rev. Genet.* **11**:79–101, 1977.
5. Brousseau, R., Scarpulla, R., Sung, W., Hsing, H. M., Narang, S. A., and Ulu, R.: Synthesis of a human insulin gene. V. Enzymatic assembly, cloning and characterization of the human proinsulin DNA. *Gene* **17**:279–289, 1982.
6. Stellwagen, R. H., and Cole, R. D.: Chromosomal proteins. *Annu. Rev. Biochem.* **38**:951–90, 1969.
7. Cattanach, B. M.: Control of chromosome inactivation. *Annu. Rev. Genet.* **9**:1–18, 1975.
8. Ris, H.: Chromosome structure. In *Chemical Basis of Heredity* (W. D. McElroy and B. Glass, eds.), John Hopkins Press, Baltimore, 1957.
9. Levine, R. P.: *Genetics*, Holt, Rheinhart and Winston, New York, 1962.
10. Ad Hoc Committee of the Environmental Mutagen Society and the Institute for Medical Research: Chromosome methodologies in mutation testing. *Toxicol. Appl. Pharmacol.* **22**:269–275, 1972.
11. Strickberger, M. W.: *Genetics*, Macmillan, New York, 1968.
12. Vogel, E., and Sobels, F. H.: The function of *Drosophila* in genetic toxicology testing. In *Chemical Mutagens: Principles and Methods for Their Detection*, Vol. 4 (A. Hollaender, ed.), Plenum Press, New York, Chapter 38, pp. 93–142, 1976.

13. Brusick, D. J.: Alterations of germ cells leading to mutagenesis and their detection. *Environ. Health Perspect.* **24**:105–112, 1978.
14. Badr, F. M., and Badr, R. S.: Studies on the mutagenic effect of contraceptive drugs. I. Induction of dominant lethal mutations in female mice. *Mutat. Res.* **26**:529, 1974.
15. Machemer, L., and Lorke, D.: Experiences with the dominant lethal test in female mice: Effects of alkylating agents and artificial sweeteners on pre-ovulatory oocyte stages. *Mutat. Res.* **29**:209, 1975.
16. Kreig, D. R.: Specificity of chemical mutagenesis. In *Progress in Nucleic Acid Research*, Vol. 2 (J. N. Davidson and W. E. Cohn, eds.), Academic Press, New York, pp. 125–68, 1963.
17. Ames, B. N., and Whitfield, H. J.: Frameshift mutagenesis in *Salmonella*. *Cold Spring Harbor Symp. Quant. Biol.* **31**:189–201, 1966.
18. Brogger, A.: The chromatid gap—A useful parameter in genotoxicology? *Cytogenet. Cell Genet.* **33**:14–19, 1982.
19. Taylor, J. H.: Sister chromatid exchanges in tritium-labeled chromosomes. *Genetics* **43**:515–529, 1958.
20. Wolff, S., and Perry, P.: Differential Giemsa staining of sister chromatids and the study of sister chromatid exchange without autoradiography. *Chromosoma* **48**:341–353, 1974.
21. Schmid, W.: Chemical mutagen testing on *in vivo* somatic mammalian cells. *Agents Actions* **3**:77–85, 1973.
22. Heddle, J.: A rapid *in vitro* test for chromosomal damage. *Mutat. Res.* **18**:187, 1973.
23. Matter, B. E., and Grauwiler, J.: Micronuclei in mouse bone marrow cells. A simple *in vivo* model for the evaluation of drug induced chromosomal aberrations. *Mutat. Res.* **23**:239–249, 1974.
24. San, R. H. C., and Stich, H. F.: DNA repair synthesis of cultured human cells as a rapid bioassay for chemical carcinogens. *Int. J. Cancer* **16**:284–291, 1975.
25. Slater, E. E., Anderson, M. D., and Rosenkranz, H. S.: Rapid detection of mutagens and carcinogens. *Cancer Res.* **31**:970–73, 1971.
26. Williams, G. M.: The detection of chemical carcinogens by unscheduled DNA synthesis in rat liver primary cell cultures. *Cancer Res.* **37**:1845–1851, 1977.
27. Samson, L., and Schwartz, J. L.: Evidence for an adaptive DNA repair pathway in CHO and human skin fibroblast cell lines. *Nature* **287**:861–863, 1980.
28. Hart, R. W., and Setlow, R. B.: Correlation between deoxyribonucleic acid excision-repair and life-span in a number of mammalian species. *Proc. Natl. Acad. Sci. USA* **71**(6):2169–2173, 1974.
29. Cerutti, P. A.: Repairable damage in DNA. In *Proceedings of ICN-UCLA Symposia on Molecular and Cellular Biology, DNA Repair Mechanisms,* (P. C. Harawalt, E. C. Friedberg, and C. F. Fox, eds.), Academic Press, New York, pp. 1–14, 1978.
30. Setlow, R. B.: Repair deficient human disorders and cancer. *Nature* **271**:713–717, 1978.
31. Ehling, U. H., Auerbeck, D., Cerutti, P. A., Friedman, J., Greim, H., Kolbye, Jr., A. C., and Mendelsohn, M. L.: Review of the evidence for the presence or absence of thresholds in the induction of genetic effects by gentoxic chemicals. *Mutat. Res.* **123**:281–341, 1983.

CHAPTER 3

# The Consequences of Genotoxic Effects in Humans and Other Mammals

## INTRODUCTION

The major factors responsible for the concern over genetic damage in human cells can be divided into two areas (Figure 3.1). The first is the concern for the protection of the human gene pool. This concern may ultimately be the most important justification for genetic testing, but it is probably the least appreciated by nongeneticists concerned with safety evaluation, because no evidence currently can be cited as proof for mutation induction in human germ cells. The second area of concern is that of oncogenesis. The intimate correlative relationship between the tumorigenic and genotoxic properties of chemicals and the identification of specific nuclear genes controlling cancer initiation (oncogenes) make genetic testing an effective screening and priority-setting technique for chemicals of unknown carcinogenic potential. This second application of genetic toxicology has been the primary driving force behind the rapid expansion of genetic toxicology as a discipline.

In assessing the consequences of genetic effects, a most important aspect is the location of the affected cell (Figure 3.2). If mutations occur in cells that are not part of the reproductive system (somatic cells), the resultant alteration affects only the individual exposed to the genotoxic agent and will not be transmitted to subsequent generations via normal reproduction. If the alteration occurs in gametes (sperm or ova) or the stem cells that give rise to gametes, the damage has the possibility to

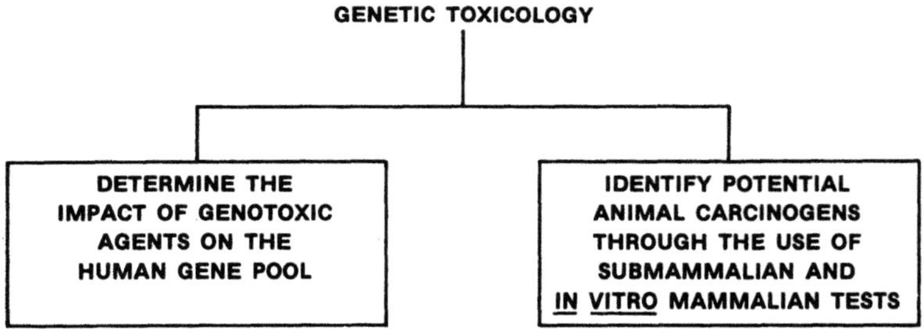


FIGURE 3.1. The two roles of genetic toxicology that have been forcing factors in its development as a discipline.

affect subsequent generations. Unfortunately, it is difficult to demonstrate a cause–effect relationship between human exposure to a mutagen and the production of mutant offspring. Using current methods, large samples of individuals and their offspring are required to show such a relationship. In addition, mutations induced in the gamates of an exposed individual may be expressed immediately if dominant, but may not be expressed for several generations if recessive. Human epidemiologic studies have been of limited value in developing a case for induced mutation because the cause–effect portions of the relationship are separated by at least one generation and involve techniques that have low powers of resolution with the sample sizes usually available.

Despite any relationship between mutagenesis and carcinogenesis, the two areas should be discussed separately, since the approaches and goals of testing for animal mutagens and for animal carcinogens are distinctly different in many respects.

## GENE POOL CONSEQUENCES

Humans, particularly those individuals in the reproductive portion of their life span, are the animal species of ultimate concern in genetic toxicology investigations. The genes that form the present generation of humans were acquired from the previous generation and the genes that we pass to our offspring will be available to form the gene pool of yet the next generation. The human gene pool is defined as the sum of the genes available in the reproductive cells of the existing population available for contribution to the next generation.

In random-mating populations, the frequency of deleterious genes occuring in one generation is related to their frequency in preceding generations. This can be illustrated using the Punnett square (Figure 3.3).

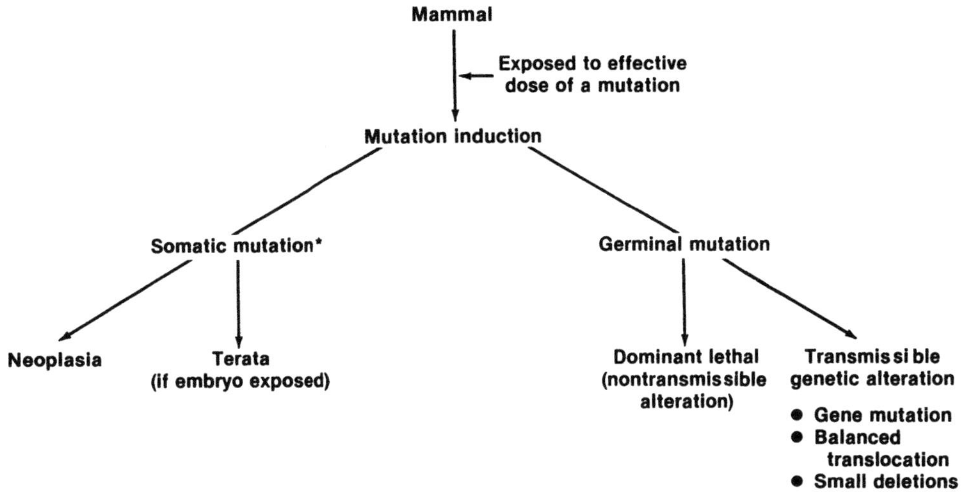


FIGURE 3.2. Possible mutagenic effects in mammals. The location of the mutagenic event is important in determining whether its effect will have an impact on the exposed individual(s) in subsequent generations. Somatic cell alterations can be detrimental to exposed individuals, whereas the alterations induced in germ cells will affect subsequent generations. Somatic mutation has also been implicated in certain types of heart disease. (From Brusick.[1])

| | | Female Gametes | |
|---|---|---|---|
| | | X (.85) | x (.15) |
| Male Gametes | X (.85) | 0.7225 | 0.1275 |
| | x (.15) | 0.1275 | 0.0225 |

FIGURE 3.3. Punnett's square. Here X is the normal gene, which resides in 85% of all individuals of the human population, and x, which is recessive to X, is the mutant form of the gene, and is carried by the remaining 15% of the population. Based on random mating, the Punnett square can be used to establish the frequency of X genotypes in the generation produced by male and female gametes of the preceding generation. Each square represents the expected frequency of gene combination. For example, approximately 98% of the next generation will appear normal for X and approximately 2% of the population will show the x trait. The proportion of individuals carrying at least one X or one x will remain constant.

Transmission of specific mutations or traits can be followed through several generations using pedigree charts of the type shown in Figure 3.4. The sum of deleterious mutations is designated as the genetic load of a population and may be thought of as a relatively constant genetic burden transmitted from generation to generation. The specific origin of genetic load is not known, but the processes of mutation (induced and spontaneous) and natural selection (preferential persistence of a given gene in random-mating populations) make important contributions to the mutational burden of a population.[2] New mutations may increase the genetic load and a substantial introduction of new mutations into the gene pool may eventually result in a generalized reduction in the reproductive capacity and overall viability of the affected species. A high genetic load in a population may lead to genetic death expressed in high levels of prereproductive death or sterility. While there is no evidence that the human genetic burden is increasing, concern for the potential of new mutations should remain a high priority.

It is often reasoned that all mutations cannot be deleterious, since without mutation no new species would ever have evolved. While it is true that mutations provide the substrate for evolutionary change, change *per se* is not necessarily beneficial. Evolutionary development is dependent on both time and random variation. The environmental scene changes with time, and biological variation results from mutation as well as new gene combinations formed by recombination during gametogenesis. If too

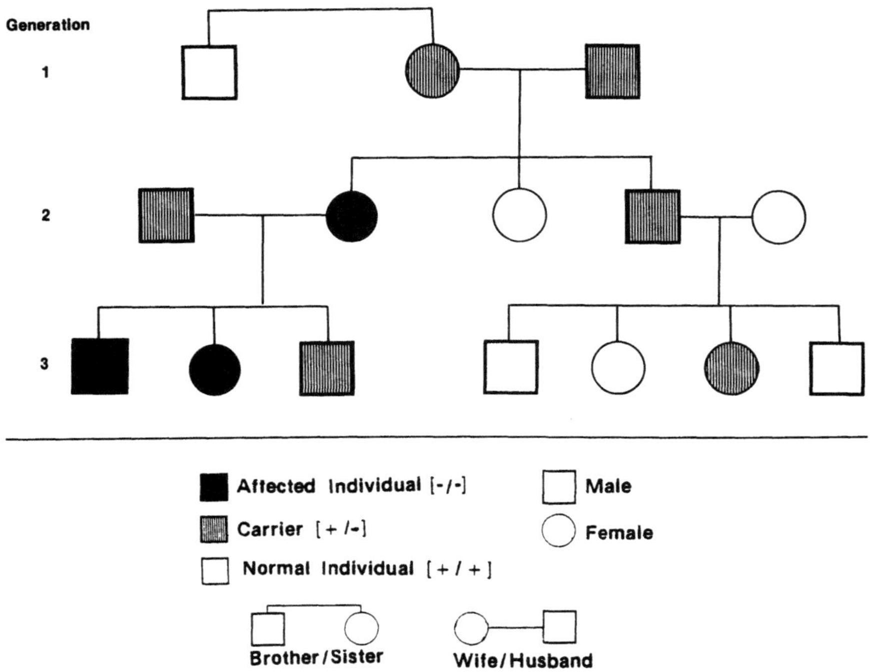


FIGURE 3.4. Pedigree analysis of an autosomal recessive genetic disease traced over three generations. In the first generation, none of the individuals are affected, although two are heterozygous carriers. The disease is expressed in one of the daughters of the two carriers. In the third generation, two individuals are affected with the disease. Construction of a family pedigree can be used to define the type of inheritance and therefore the nature of the genetic lesion.

much variation occurs in a short time span, or if too much time elapses without genetic change, serious biological problems may arise for a species and threaten its existence. A mutation that might reduce the viability of a species living under one set of environmental conditions might result in a trait advantageous for other conditions.

Changes in the composition of the gene complement in the present environment would probably result in genetic dysfunction, leading to disease or death. While discussions of beneficial or deleterious mutations can thus be best interpreted only when the natural selection process is fully understood, most authorities believe that a rapid increase in mutation rate resulting from exposure to environmental mutagens will almost certainly result in an increased genetic burden for the human species and a corresponding increase in health care costs associated with genetic dysfunction in subsequent generations. It is estimated that direct damage to the genetic material or induced modulation of its function forms the basis for a substantial portion of human health care expense.[3,4] The goal of genetic toxicology is to protect the integrity of the gene pool inherited from our ancestors and to pass to our heirs healthy sets of gametes.

## GENOMIC TARGET IN HUMANS CELLS

There are substantial differences between DNA size and function among the experimental models used to measure the mutagenic activity of chemicals. The human genome (total complement of genetic information) consists of 23 pairs of chromosomes. The total number of genes present on those chromosomes has been estimated at $3 \times 10^4$ to $1 \times 10^5$, comprised of $3 \times 10^9$ base pairs (in a haploid chromosome set).

Not all genes in the human genome are involved in transcription; in fact, some estimates put the number of active genes at only 2% of the total DNA. Extensive regions of repetitive DNA sequences exist, making estimates of critical target size relatively uncertain.[5] It appears that large areas within the human genome are unaffected by gene mutation or chromosomal rearrangements. While the genomes of animal models also share this phenomenon of extensive repetitive DNA sequences, other organisms, such as the bacteria used in microbial models, lack this genomic expansion and may be inadequate for use in quantitative risk analyses.

## THE DIRECT EFFECT OF MUTAGENS ON THE TRANSMISSIBLE GENE POOL

In spite of the apparent refractile potential of the human genome for mutagenic change and scientists' inability to demonstrate induced mutagenic damage in humans, a sizable genetic burden exists in the human population. Estimates of the human genetic load range from 1 to 2% among new births and include chromosome abnormalities and autosomal dominant, recessive, and X-linked gene mutations.

The latest compilation of identified genes in humans lists 1611 actually confirmed and 1692 partially confirmed, for a total of 3303. Table 3.1 gives the distribution of the entire 3303 loci across the various inheritance categories.

One might conclude that with a normal spontaneous mutation incidence and chemical-induced mutations occurring from environmental exposure the human gene pool will rapidly deteriorate. There are several reasons why this will not happen. There is a balance between spontaneous forward and reverse mutation rates (i.e., the probability that a normal

TABLE 3.1.
Distribution of Identified Human Genes by Type of Inheritance

| Inheritance | Number of identified loci | |
|---|---|---|
| | Confirmed | Partially confirmed |
| Autosomal dominant | 904 | 880 |
| Autosomal recessive | 591 | 692 |
| X-linked | 116 | 120 |
| Total | 1611 | 1692 |
| Grand total | 3303 | |

[a] From Matsunaga.[6]

gene will mutate and the probability that a mutant form of the gene will backmutate to a normal form). It is also possible that, for a variety of reasons (Table 3.2), many new mutations induced by genotoxic agents are not transmitted via gametes. Even if a mutation is transmitted, its chance of survival in the gene pool is low unless there is selective pressure to maintain the trait.

A mutation would have a high probability for fixation when the mutant gene confers some selective advantage to the heterozygous carrier. Evidence for this type of heterozygous selection has been reported for some human diseases. A notable example is sickle-cell (designated by S) disease, in which the unaffected Ss heterozygote carrier appears to be more resistant to malaria parasites than are homozygous SS nonmutant individuals. Thus, under the proper set of environmental conditions (i.e., high levels of malaria), the Ss heterozygous carrier is more likely to survive than the normal SS individual, while homozygous recessive individuals affected by sickle-cell disease (ss) may be severely debilitated or die.

One might expect that if a homozygous trait such as sickle-cell disease (ss) were 100% lethal before the reproductive age of the affected individual, the disease would disappear. This is not the case, since recessive lethal traits are in most respects masked when in the heterozygous (normal:mutant) combination. The incidence of homozygous (ss) individuals in the population is dependent on the frequency of heterozygous (Ss) carriers in the population (see Figure 3.3 for explanation). Most humans are carriers for one or more recessive lethal traits.

Genetic alterations fall under several categories (Table 3.3). Among these categories are alterations that result from the gain or loss of chromosomes, portions of chromosomes, or entire sets of chromosomes. The most common chromosome anomalies occur in the sex-chromosome pair

TABLE 3.2.
Factors Involved in Preventing the Establishment of a Single New Mutation in the Gene Pool[a,b]

| Critical step | Reason for loss of the mutation |
|---|---|
| Meiosis[c] | Mutations induced in stem or gonial cells must be able to traverse events in meiosis such as DNA/chromosome replication; many types of chromosome aberrations are lethal when subjected to the processes of mitosis and cell division |
| Ejaculation | Assuming that the mutation is compatible with meiosis and resides in a mature sperm, it may be lost in the seminal fluid and never reach the region of fertilization |
| Fertilization | From the millions of cells in an ejaculation, the number of sperm involved in fertilization is low (1–20, depending on the species); thus, most sperm from an ejaculation will be lost; if the new mutation is not in a gonial stem cell, it will rapidly be diluted by nonmutant sperm and have a low probability for involvement in fertilization |
| Implantation | Mutations that survive meiosis and reside in a sperm that fertilizes an egg may be lost if their expression interferes with zygote implantation or embryonic development; early death and resorption will occur; this would probably be true only for dominant mutations; postfertilization repair, known to occur in mammalian zygotes, could affect expression of mutation |
| Embryonic development | Mutations may affect processes of cellular differentiation or hormonal regulation necessary for maintenance of the embryo and fetus; expression of these mutations may result in early abortion |
| Early death | Mutations that affect the prereproductive survival of the progeny will be eliminated by early death of the affected progeny before transmission and fixation of the mutation in the gene pool |
| Transmission capacity | Assuming that an $F_1$ individual is heterozygous (+/−) and mates with a normal (+/−) individual, the probability of fixation of the new mutation is a function of the number of their offspring; sterility is also a factor |

[a] This logic applies to a single new mutation in a premeiotic or postmeiotic cell. It does not consider the total number of mutations per cell or total number of mutations per cell population.

[b] The events shown in this table make it appear as if no new mutations would be transmitted from one generation to the next. The events were developed, however, by looking at a single mutation in a single maturing gamete. When an animal is exposed to a mutagen, stem cells, all developing gametes, and all genes within the gametes are potential targets. Thus, for a potent mutagen it may be possible that at least one new mutation will be induced in each maturing gamete, ensuring that some genetic change will be transmitted to the progeny.

[c] Applicable only to mutations in premeiotic cells. All remaining steps are applicable to either pre- or postmeiotic cells.

TABLE 3.3.
Examples of Genetic Disorders in Humans

| Disorder | Estimated frequency/$10^3$ population[a] | Typical examples |
|---|---|---|
| Chromosomal abnormalities | 6.86 | Down's syndrome (trisomy 21) |
| | | Klinefelter's syndrome (XXY) |
| | | Turner's syndrome (XO) |
| | | Cri du chat (deletion of chromosome) |
| | | Numerous other trisomies |
| | | XYY |
| Dominant mutation | 1.85–2.64 | Familial polyposis (AD) |
| | | Neurofibromatosis (AD) |
| | | Huntington's chorea (AD) |
| | | Hepatic porphyria (AD) |
| | | Crouzon's craniofacial dysostosis (AD) |
| | | Achondroplasia dwarfism (AD) |
| | | Retinoblastoma (AD) |
| | | Anitidia (AD) |
| | | Chondrodystrophy (AD) |
| Recessive mutation | 2.23–2.54, 0.78–1.99 | Xeroderma pigmentosum (AR[c]) |
| | | Duchenne muscular dystrophy (XR[d]) |
| | | Hemophilia (XR) |
| | | Lesch–Nyhan syndrome (XR) |
| | | Sickle-cell disease (AR) |
| | | Galactosemia (AR) |
| | | Phenylketonuria (AR) |
| | | Diabetes mellitus (AR?) |
| | | Fanconi's syndrome (AR) |
| | | Albinism (AR) |
| | | Cystic fibrosis (AR) |
| Polygenic (complex inheritance) | 26.00–32.00 | Cleft lip |
| | | Anencephaly |
| | | Spina bifida |
| | | Clubfoot |
| | | Idiopathic epilepsy |
| | | Congenital heart defects |

[a] From Ref. 7. AD, Autosomal dominant. AR, Autosomal recessive. XR, X-linked recessive.

and result in XO (Turner's syndrome), XXY (Klinefelter's syndrome), or XYY individuals. Chromosome anomalies such as deletions or rearrangements also produce characteristic human disorders. Chromosome aberrations may be either balanced or unbalanced. Cells with balanced aberrations contain an active complement with rearrangements, whereas unbalanced aberrations show net gain or loss with little or no chance for a normal cell division process. Balanced events may be compatible with survival.[8] Table 3.4 shows that the frequency of different types of aberrations varies between live births and spontaneous abortions.

The mechanism believed to produce changes in chromosome num-

TABLE 3.4.
Relative Frequencies of Chromosome Abnormalities in Human Live Births and Spontaneous Abortions[a]

| Observed relative frequency | Live births | Spontaneous abortions |
|---|---|---|
| Highest | Structural | Trisomy |
| | Monosomy | Monosomy |
| | Trisomy | Polyploid |
| Lowest | Polyploid | Structural |

[a] The data illustrate the effects of chromosome alterations on organism lethality. Chromosomal changes involving sets or whole chromosomes are generally less compatible with life than breaks and gaps and are frequently observed in abortions. Structural aberrations seldom lead to spontaneous abortion.

bers is nondisjunction, that is, the failure of a chromsome pair to separate during one of the two meiotic reduction divisions (Figure 3.5). Nondisjunction may occur in sex chromosomes and in non-sex chromosomes (autosomes). Down's syndrome results from the presence of an extra chromosome (number 21).[9] The addition of one chromosome produces extensive physiologic and psychological changes in the affected individuals, including mental retardation, sterility, and shortened life expectancy. Autosomal nondisjunction is associated with late pregnancy in women,[9] but genotoxic agents have also been identified with the production of nondisjunction in submammalian experimental systems.[10] It is currently believed that most nondisjunction in mammals leading to trisomy of autosomes or sex chromosomes is spontaneous, and considerable research is needed to establish whether or not chemicals can induce this type of genetic alteration. There is no reliable experimental data base or assay method with which to address this issue effectively.

Chemicals that preferentially produce chromosome alterations without the concomitant induction of mutations at the gene level are designated clastogens. Most aberrations are lethal to the cell, but certain types, such as balanced translocations and small deletions, can be transmitted through the germ lines.[8] For instance, the mutations forming the second and third categories in Table 3.1 are gene mutations that have a high level of transmissibility. They affect only a single gene and do not interfere with chromosome replication or division. Figures 3.6–3.8 illustrate the impact on the gene pool of the three types of gene mutations.

A dominant mutation is expressed in the first generation following its production. If the dominant mutation is lethal, it will be expressed early in the zygote or developing embryo. Even if the sperm is capable

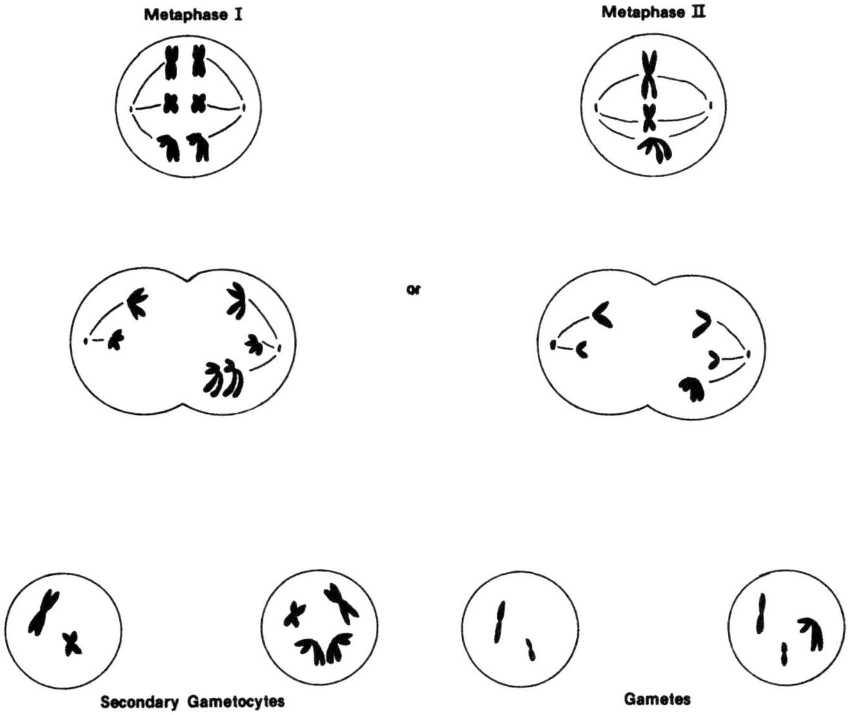


FIGURE 3.5. Illustration of the process of nondisjunction occurring at either metaphase I or metaphase II of meiosis. In both situations shown, one cell receives two chromosomes of the same pair while the sister cell loses an entire chromosome. Nondisjunction need not always affect an entire chromosome, but occasionally small pieces are involved. Nondisjunction generally leads to serious genetic problems in humans. Trisomy 21 (Down's syndrome) is a typical example.

of fertilizing the ovum, the resultant zygote will die before a mature fetus is formed. Dominant lethal mutations, therefore, have little impact on the gene pool and are self-limiting events.

Dominant viable mutations (Figure 3.6) are also expressed immediately, but they do not result in embryonic death and can be observed in the first or $F_1$ generation. These mutations will have an impact on the gene pool, but since both their presence and transmission are readily identifiable, parents can decide in advance if the risk of birth is acceptable. Once detected, dominant viable and sex-linked traits in populations can easily be followed. While screening for a group of 10–12 such traits might be useful for monitoring the impact of mutagens on exposed populations, the large populations required for accurate estimates make the prospects for this proposal very unlikely.

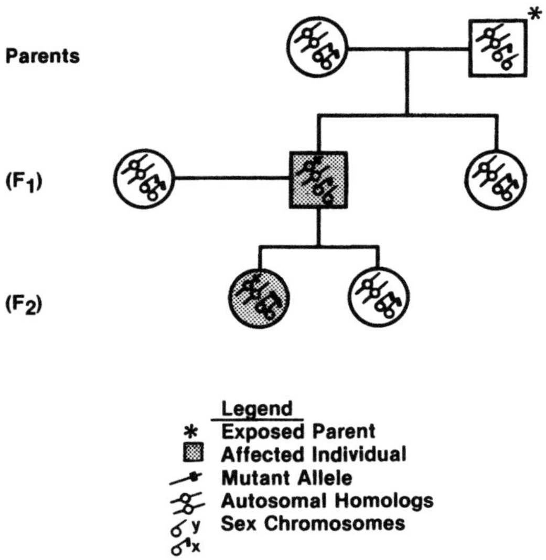


FIGURE 3.6. New autosomal dominant mutation (mutation shows a high degree of penetrance and expression). The mutations can be induced in gametes of either parent and the traits (phenotype) will be expressed in the first generation ($F_1$). The phenotype can be expressed in offspring of either sex. Only one mutant allele is required for mutant expression. If an affected individual marries a normal individual and they have children, there is a probability of 0.5 that each offspring will acquire the mutant gene.

Recessive autosomal mutations are not expressed unless the affected individual receives mutant genes from both parents (Figure 3.7). Two heterozygous carriers will appear normal for the trait and not realize they carry the recessive mutant gene. Upon mating, their probability of producing an offspring showing the trait is 1 in 4. Having an affected offspring and realizing that they are both heterozygous for a trait, parents now have important information on which to base future birth decisions.

Semidominant alleles and X-chromosome (sex-linked) mutant alleles also contribute to the genetic burden. Both alterations are expressed more rapidly than recessive mutations, but, unlike dominant alleles, generally not in the first generation. Their expression is also restricted to specific conditions (e.g., sex of offspring, as in Figure 3.8).

Because of the homozygosity requirement, new recessive mutations introduced into the gene pool will probably not be expressed for several generations. Accumulation of recessive mutations in the gene pool as a result of exposure to environmental mutagens could go undetected for generations and may result in a serious genetic burden, and the problem would be more serious than just the costs associated with health care for affected individuals. The induced mutations would now become a permanent part of the genetic load, with the consequences being higher incidence of disease in the population due to the higher frequency of heterozygous carriers. These consequences will continue to exist even though the causative agent inducing the mutations may have been removed from the environment.

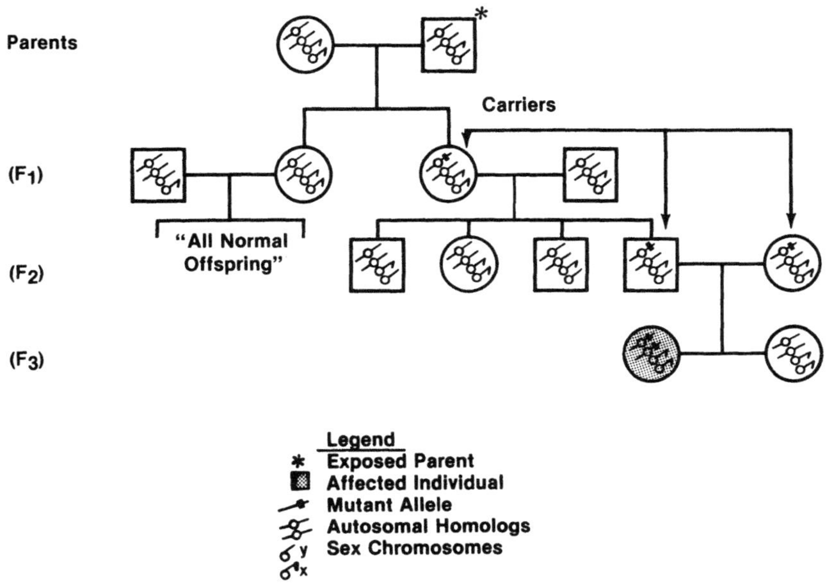


FIGURE 3.7. New autosomal recessive mutation. The mutation can be induced in gametes of either parent; however, the phenotype will not be expressed in the first generation ($F_1$) offspring of either sex. The offspring identified as the carrier is heterozygous for the allele and will only contribute the gene to the next generation (i.e., $F_2$) if married to an individual with a normal allele at that site. If two carriers heterozygous at the same site (as shown in the $F_2$ mating), marry, a probability of 0.25 exists that each offspring will acquire the mutant allele from both parents (homozygous) and express the mutant phenotype.

The most complex situation occurs when hereditary traits are influenced by several recessive gene sets. These traits, called polygenic characters, require a specific combination of several genes to produce the expression of a particular trait. Many new mutations could be induced in this set of genes without ever being noticed. Polygenic traits are discussed in more detail in Chapter 10.

Assuming that a new mutation is transmitted and established in a population, the expression of the mutant phenotype depends on several factors not well understood. Various secondary factors can influence the level of expression for any mutant gene. The expression may vary among individuals (dependent on genetic background) and with sex and age. The frequency with which the trait is expressed is designated penetrance, and the level to which the trait is expressed is called its expressivity. Thus, mutations with complete expressivity will show the full set of features

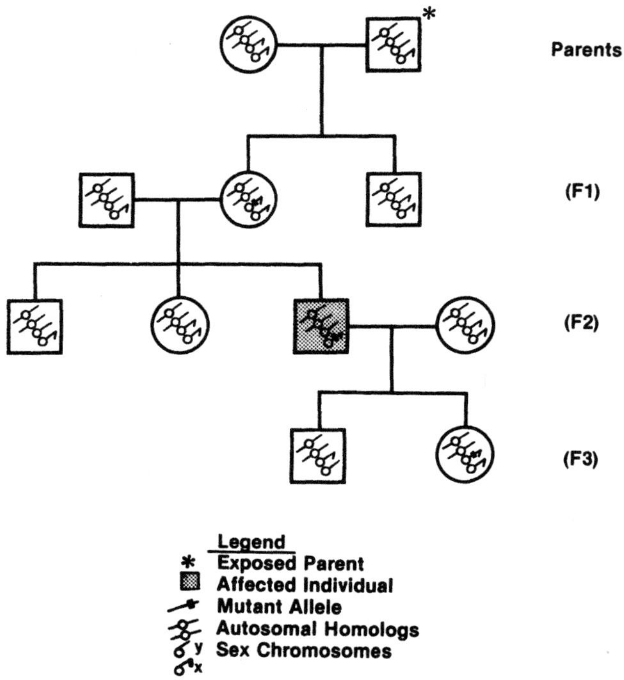


FIGURE 3.8. New X-linked recessive mutation. New X-linked mutations would only be expressed in male offspring. Female progeny carrying a single mutation would show an inheritance similar to an autosomal recessive mutation. Only females carrying two mutations will exhibit the trait (phenotype). Male offspring cannot inherit the trait from an affected father ($F_3$).

associated with the alteration, whereas mutants with incomplete expressivity will result in some fraction of the characteristic features is affected individuals. The penetrance and expressivity of a particular mutation cannot be assessed in advance of its expression in a given individual except in a most general way. For example, the severity of a particular genetic disease may range from subclinical to highly debilitating even among members of the same family.

A certain perspective must be maintained when assessing the impact of environmental mutagens on the human gene pool. This perspective should encompass the impact of modern medicine on maintaining deleterious genes in the population. With the aid of newborn monitoring and appropriate therapy, the lethality of many genetic diseases that would otherwise have resulted in prereproductive death can now be avoided, permitting more "genetically defective" individuals to lead normal lives, which includes producing offspring. Without doubt, therapeutic intervention of genetic disorders will contribute more to a higher genetic load via proliferation of preexisting mutations than induction of new mutations via environmental mutagens.

## THE RELATIONSHIP OF GENOTOXIC EFFECTS TO OTHER TOXICOLOGIC PHENOMENA

Because of the fundamental role that genes play in all aspects of living organisms, the concept that altered genes will lead to various disease states has considerable merit.[3] Direct evidence is available for diseases inherited as single-gene traits (Table 3.3). In these examples, mutant genes result in defects in mental capacity, development of sexual characteristics, vision, bone development, skin pigmentation, blood cell formation, muscle development, hormone balance, enzymatic processes, and embryonic development.

Other types of toxicologic endpoints appear to be under some type of genetic control, but the evidence is only indirect. Among these endpoints are:

1. Oncogenesis (certain forms such as retinoblastoma and familial polyposis are clearly inherited)[11]
2. Teratogenesis[12]
3. Sterility or semisterility[13]
4. Heart disease[14]
5. Aging[4]

It is known that genotoxic agents can produce alterations that interfere with cell differentiation and produce terata; however, a high proportion of teratogenic agents are not mutagenic and act through nongenetic mechanisms.[13] Consequently, short-term mutagenicity tests cannot be relied upon to define teratogenic agents. Recently a range of *in vitro* cell culture, embryo culture, and invertebrate models relying upon the detection of changes in molecular events such as protein synthesis or cell organization have been proposed as predictive tests for teratogenicity. One *in vitro* method, which assesses the synthesis of sulfated proteoglycans in cultured embryo cells following exposure to putative teratogens, has shown a better than 90% predictability for a diverse group of rodent teratogens and nonteratogens.[15]

The primary emphasis placed on genetic testing, however, has been its ability to identify potential carcinogens. Several factors have been instrumental in increasing the emphasis on short-term carcinogenicity tests. These include the high cost and the long duration required to perform rodent oncogenicity studies, and the desire to determine the mechanism of tumor production in rodent models. Costs for typical chronic oncogenicity studies in mice and rats have increased dramatically over the past 10 years, making the decision to perform such studies on a particular chemical a major financial and logistic undertaking. Thus, any information that can be obtained in advance regarding the likely outcome of such studies is extremely valuable in go–no-go decision-making processes.

A somatic mutation theory for the etiology of cancer was proposed by Bauer in 1928.[16] The data base was insufficient to support such a theory through the 1950s[17] and even the late 1960s.[18] The type of correlative data base to support a mutation step in the induction of tumors was not available until the early 1970s, when microbial mutation assays were coupled with microsomal metabolic activation (S9 mix) systems.[19] Many of the experimental animal carcinogens are biotransformed *in vivo* into biologically active agents and bacterial or *in vitro* mammalian cell assays cannot carry out such transformation without an exogenous metabolic source. Early analyses of animal carcinogens in S9 microsome enzyme-supplemented microbial tests produced highly suggestive evidence that most (85–90%) animal carcinogens were also mutagens.[20] The degree of correlation between these two phenomena appeared to be indicative of an intimate functional relationship, and the probability that a genotoxic event was essential to the oncognic process gained considerable support.[21–25]

Based on this relationship, a generalized hypothesis associating mutagenic and carcinogenic mechanisms was proposed (Figure 3.9). Chemicals that fit this scheme have been referred to as mutagenic carcinogens, DNA carcinogens, or initiating carcinogens, implying that their mechanism of action involves genotoxic events.

There were numerous problems associated with trying to use correlative evidence as a basis of proof. While the correlation between animal carcinogens and their responses in *in vitro* assays is not trivial, the predictive values may range from 0.60 to 0.95, depending on the composition of the group of chemicals employed.[26] Biasing the group with aromatic amines, polycyclic hydrocarbons, nitrosamines, and direct-acting alkylating agents will produce an exaggerated relationship; whereas the inclusion of a high proportion of metal carcinogens, hormones, halogenated organic molecules, and aromatic solvents will result in a reduction of the correlation. Thus, correlations can be manipulated and probably represent the least convincing evidence for the carcinogen/mutagen relationship.

Table 3.5, for example, shows a large number of the published and unpublished comparative results for the Ames assay. The sensitivity (number of carcinogens that are also mutagens) and specificity (number

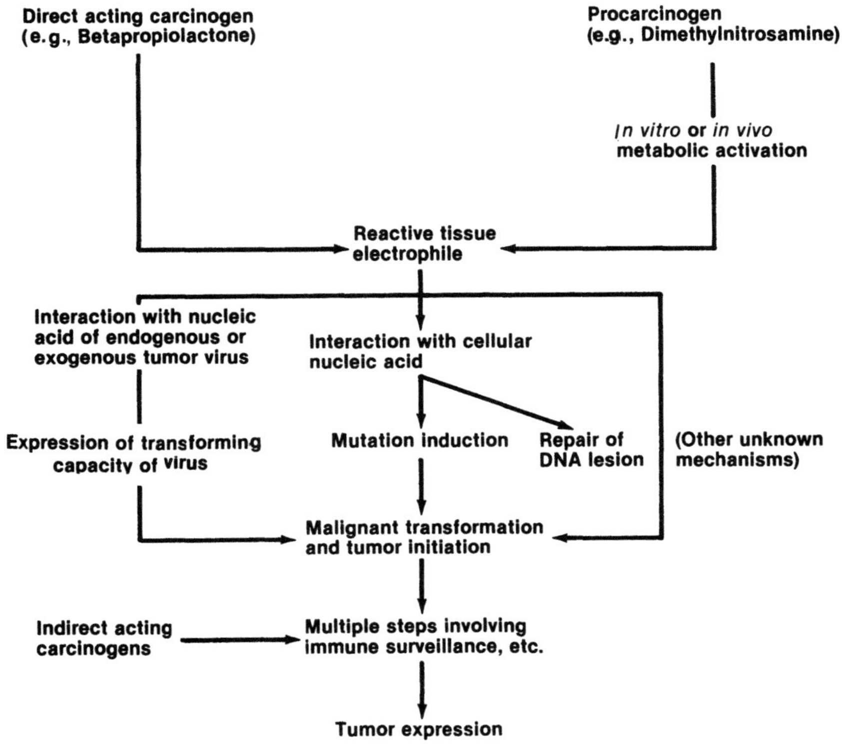


FIGURE 3.9. Proposed relationship between carcinogenicity and mutagenicity. This early model accounts for many facets of what is known about tumor production in mammals. While there is significant indirect support for this model, formal proof is not yet available. There are a number of agents known to influence the expression of malignancy in cells once the initiation process has occurred. However, these agents are not capable of producing the initiation event.

of noncarcinogens that are nonmutagens) vary widely. The variability appears to be related to three factors:

1. *Method by which the chemical sample used in the comparison was selected.* This factor appears to be extremely important in determining the outcome of the comparative exercise. Studies in which the investigator selected the chemicals to be included in the sample show significantly higher correlations than do studies using chemical samples that were not influenced by the investigator.

2. *Tests on coded and uncoded samples.* The correlations obtained from tests with coded samples were lower than those found in studies in which the investigators knew the identity of the materials at the time of testing or at the time of data interpretation.

TABLE 3.5.
Assay Performance of the Ames *Salmonella* Reverse-Mutation Assay[a]

| Ames assay assessment | Year reported | Chemicals in the sample | Sensitivity (%) | Specificity (%) | Predictive value (%) | Basis of chemical selection |
|---|---|---|---|---|---|---|
| McCann *et al.*[28] | 1975 | 300 | 90[b] | 87[b] | 92 | Investigator selected |
| Sugimura *et al.*[29] | 1976 | 241 | 93[b] | 77[b] | 87 | Investigator selected |
| Purchase *et al.*[30] | 1978 | 120 | 91[b] | 93[b] | 93 | Investigator selected |
| Simmon[31] | 1979 | 92 | 64 | 78 | 90 | Committee selected |
| Andersen *et al.*[32] | 1979 | 221 | 90 | 100 | 100 | Investigator selected |
| Rinkus and Legator[33] | 1979 | 271 | 77[c] | NA[d] | — | Literature data base |
| Bartsch *et al.*[34] | 1980 | 89 | 76[d] | 57[b] | 95 | Investigator selected |
| Bridges *et al.*[35] | 1981 | 42 | 45–70 | 60–80 | — | Investigator selected |
| NIEHS/NTP[36] | 1982 | 218 | 54 | — | 76 | Selection committee |
| ICPEMC[37] | 1982 | 51 | 64 | 35 | 58 | Animal germ cell mutagens |
| NCI Comparative[38] | 1982 | 58 | 78 | 67 | 81 | Selection committee |
| EPA GeneTox[39] | 1984 | 245 | 80 | 42 | — | Literature data base |

[a] All values reported are approximate and based upon evaluations of results by the authors of the reports or individuals giving their personal communications. This table is adapted from Brusick.[26]
[b] From I. F. H. Purchase.
[c] Reanalysis of the report by Ames and McCann showed an 82% correlation.[27]
[d] NA, Not applicable.

3. *Tests conducted in a single laboratory or in multiple laboratories.* Test results from multiple-laboratory data tend to produce lower correlation values because of protocol variations and the need for consensus responses. A special variation of this problem was inherent in those studies that relied on published data, because few reports itemize the responses from all trials conducted during the evaluation. Consequently, the true reproducibility of a reported response may not be reflected in the final manuscript.

## FALSE-NEGATIVE AND -POSITIVE RESPONSES

Based on the evidence linking carcinogens to genotoxic lesions, one might expect a much greater degree of correlation between short-term mutagenicity test results and cancer in animals than is derived from correlation studies. The poor concordance is generally attributed to unacceptably high false-positive and false-negative responses in the predictive tests.

A counterargument proposes that both the rodent assay and the short-term tests experience false responses and that the poor concordance is not due to inadequacies of the short-term tests, but rest with uncertainties about the animal cancer data used to "validate" the short-term bioassays. Responses in one animal species generally cannot be extrapolated, either quantitatively or qualitatively, to a different species or even a different strain of the same species.[40] When concordance does not exist among animal models, it should not be surprising that variable performance against predictive tests is encountered. When conventional animal models (mouse and rat) exhibit both positive and negative carcinogenic responses to a chemical, the critical question is: which of the two species most accurately predicts the human responsiveness and does the short-term test result match the result from the most accurate species?

Because of conflicting animal responses and questions on the significance of some current animal studies, the use of animal data for an assessment of the reliability of mutagenicity tests has been reduced. Human carcinogenesis should be the true standard to determine the validity of a predictive test; however, the human carcinogen data base consists of only agents identified as positive. By using the human carcinogens to validate tests, one finds that a combination of one or two short-term bioassays is as predictive for positive results as are one or two rodent species (Table 3.6). This analysis fails to determine specifity and is thus of limited value in evaluating test performance.

A discussion of false results is confusing at best since the basis for it assumes that the truth can be clearly defined. Unfortunately, in the

TABLE 3.6
Responses of Human Carcinogens in Rodent Tumor Models and *in Vitro* Short-Term Assays[a]

| Group 1 chemical | *In vivo* tumor assay | | | | Ames | Mammalian cell mutation assays | Cell transformation | Cytogenetics | | |
|---|---|---|---|---|---|---|---|---|---|---|
| | Human | Mouse | Rat | Other | | | | ABS | SCE | UDS |
| 1. 4-Aminobiphenyl | + | + | + | + | + | ND | + | ND | ND | + |
| 2. Analgesic mixtures containing phenacetin | + | – | + | | + | ND | ND | + | + | ND |
| 3. Arsenic and arsenic compounds | + | – | – | | – | + | + | + | + | ND |
| 4. Asbestos | + | + | + | + | – | (+) | + | + | ND | ND |
| 5. Azathioprine | + | (+) | (+) | | + | ND | ND | + | ND | ND |
| 6. Benzene | + | – | – | | – | – | – | + | ND | ND |
| 7. Benzidine | + | ND | + | + | + | + | + | + | + | + |
| 8. *N,N*-Bis(2-chloroethyl-2) naphthylamine (chlornaphazine) | + | + | + | | + | ND | + | ND | ND | ND |
| 9. Bis(chloromethyl)ether and chloromethyl methyl ether | + | + | + | | + | ND | + | + | + | ND |
| 10. Myleran | + | (+) | – | | + | ND | + | + | + | ND |
| 11. Certain combined lymphoma chemotherapy (MOPP) | + | ND | ND | | ND | ND | ND | ND | ND | ND |
| 12. Chlorambucil | + | + | + | | + | ND | ND | + | ND | ND |
| 13. Chromium and chromium compounds | + | – | + | | + | + | + | + | + | + |
| 14. Conjugated estrogens | + | ND | – | | – | ND | ND | – | ND | ND |
| 15. Cyclophosphamide | + | + | + | | + | + | + | + | + | ND |
| 16. Diethylstilbestrol | + | + | + | + | – | (+) | + | + | + | – |
| 17. Melphalan | + | + | + | | + | + | + | + | + | + |
| 18. Methoxysalen plus UV light (PUVA) | + | + | ND | | + | + | + | + | + | ND |
| 19. Mustard gas | + | + | ND | | + | + | ND | + | ND | ND |
| 20. 2-Naphthylamine | + | + | – | + | + | + | + | + | + | ND |
| 21. Soots, tars, and oils | + | + | – | | + | + | + | + | + | + |
| 22. Treosulphan | + | ND | ND | ND | ND | ND | ND | ND | ND | ND |
| 23. Vinyl chloride | + | + | + | + | + | + | ND | + | + | ND |

[a] ND, No data available. ( ), Limited data. Data obtained from Refs. 41 and 42.

case of carcinogen prediction from short-term testing, false and true responses are only relative and will tend to change with time and the accumulation of greater knowledge. The arguments against the use of genetic toxicology because of too many "false positives" are probably invalid, since defining the true potential of noncarcinogenic mutagens to increase the tumor rate in exposed humans cannot be accomplished by animal models. For example, it would require a population of 10,000 rodents to detect a 1% increase in tumor incidence with the bioassay approach using the strains of animals typically employed. Consequently, weak carcinogens may not show significant tumorigenesis in the rodent bioassay.

Such agents as AF-2, ethylene dibromide, or the flame retardant Tris might still be prevalent in the human environment if genetic toxicology results had been ignored because they were likely to represent false-positive responses.

## Promoting Agents as False Negatives

Some agents induce tumors in rodents apparently via mechanisms without the involvement of genotoxicity. In structured two-stage experiments involving initiation followed by repeated application of the test agent, many of these agents enhance the tumor development substantially. Chemicals that enhance tumor development in such experimental models are classified as promoting agents. Promoters have historically shown little or no genotoxic activity in short-term assays and were therefore considered to be false negatives. Recently some promoting agents (TPA and teliocidin) have shown weak genotoxicity or have shown an ability to induce peroxisome formation.[43] Peroxisome induction leads to higher levels of peroxides (e.g., $H_2O_2$) and other free radicals in the target tissues. Hydrogen peroxide and other oxygen radicals are genotoxic in many systems.[44] Thus, the presumption that promoters induce tumors without genotoxicity involvement may not be entirely correct.

A separate group of chemicals may promote tumor induction without clear genotoxicity involvement. Chemicals that often appear in this group are the polychlorinated organics, phenobarbital, sodium saccharin, and nitrilotriacetic acid (NTA). Genetic toxicology assessment of these agents has consistently failed to show reproducible evidence of DNA-damaging activity; this group of agents does have one common feature, in that they or their salts appear to be directly involved in tissue hyperplasia.[45] It is this biological feature of these agents that appears to be associated with tumor development. Cell turnover may result in the expression of inherited preinitiated cells. Numerous examples can be cited for inherited predisposition to specific tumors in strains of mice (e.g., liver tumors <75%

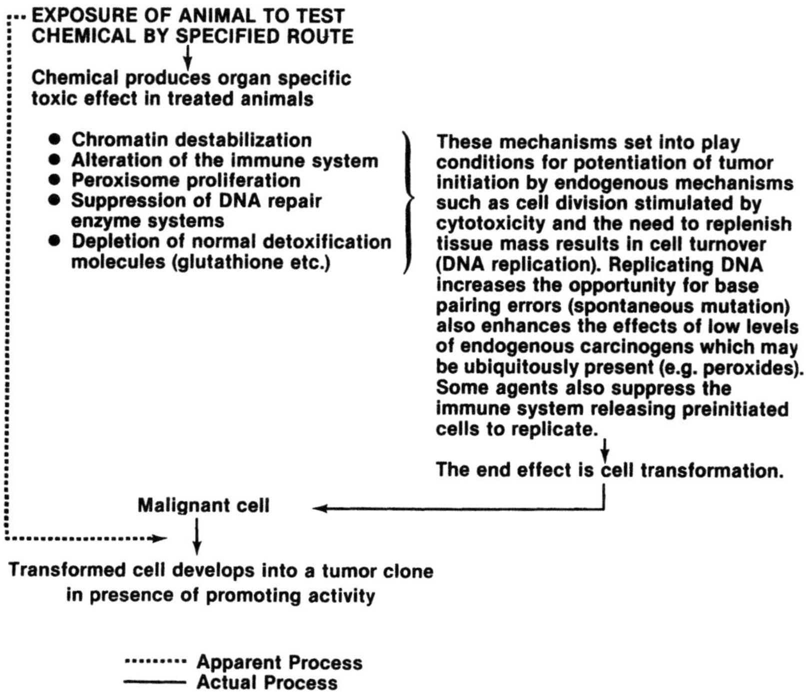


FIGURE 3.10. Hypothetical steps leading to tumor initiation by indirect carcinogens. Many agents, while not initiating cell transformation, can lead to the enhancement or expression of transformed cells. Some of the possible mechanisms are shown. (From Brusick.[21])

in male B6C3F1 mice) and rats (e.g., testicular interstitial cell tumors <5% in male Fischer rats; mammary carcinomas <5% in female Fischer rats).

A secondary implication of an acceptance of two mechanisms for tumor induction (one genotoxic and one nongenotoxic) would be to suggest a true no-effect or threshold level for nonmutagenic carcinogens.[46] The no-effect level should be the level of exposure below which no toxic

TABLE 3.7
Presumed Mechanisms of Nongenotoxic Carcinogens

| |
|---|
| Selective stimulation or proliferation of preneoplastic lesions |
| Activation or promotion of oncogene expression |
| Inhibition of gap junction proteins that maintain cell stabilization |

effects associated with cellular stimulation are observed. At present there is not much published data demonstrating the specific points identified in Figure 3.10; however, the carcinogenic mechanisms of agents such as nitrilotriacetic acid, sodium saccharin, DDT, and other nonmutagenic carcinogens appear, in general, also to fit the scheme. Table 3.7 identifies three potential mechanisms for nongenotoxic carcinogens.

## RELATIONSHIP OF POTENCY BETWEEN MUTAGENICITY ASSAYS AND *IN VIVO* ONCOGENICITY

Several investigators have begun to look at the potency ranges for mutagenic and carcinogenic properties of mutagenic carcinogens to determine whether they are similar or dissimilar. If similar, then *in vitro* mutagenicity studies might be useful for estimating the *in vivo* potency of suspect carcinogens. Early reports by Meselson and Russell[47] and Clive[48,49] indicated that the activity of both mutagens and carcinogens covered a $10^6$-fold range and comparisons produced a roughly linear relationship between mutagen potency and an expression of carcinogenic potency (Figure 3.11). Considerable debate was stimulated by these re-

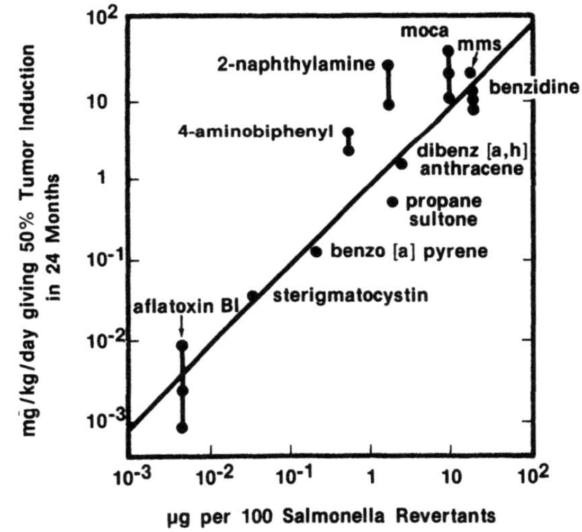


FIGURE 3.11. Relationship between potency of mutation and cancer. Several investigators have attempted to relate potency in short-term tests with potency in animal studies. While this concept is potentially very interesting, technical problems will probably prevent the general use of such comparisons. mms, Methylmethane sulforate; moca, 4'4'-methyene bis(2-chloraniline). (From Meselson and Russell.[47])

ports and numerous investigators attempted to verify this relationship with a broader range of agents than described for the Ames and mouse lymphoma assays. For some classes of chemicals in some tests, the relationship between mutagenic activity (revertants/nmole or mutant frequency/dose unit) and cancer ($TD_{50}$) holds reasonably well. Polycyclic hydrocarbons make up one such class. Other classes of chemicals, such as the nitrosamines and aromatic amines, do not show a relationship between genotoxic and carcinogenic potencies. These carcinogens show relatively high degrees of organ and species specificity, which might account for the inconsistent activities. It appears that potency is a multifaceted property of a chemical and more sophisticated approaches are required in order to provide an *in vitro* surrogate.

## REFERENCES

1. Brusick, D. J.: Alterations of germ cells leading to mutagenesis and their detection. *Environ. Health Perspect.* **24**:105–112, 1979.
2. Van Valen, L.: Selection in natural populations. III. Measurement and estimation. *Evolution* **19**:514–528, 1965.
3. Ames, B. N.: Identifying environmental chemicals causing mutations and cancer. *Science* **204**:587–593, 1979.
4. Burnet, F. M.: *Intrinsic Mutagenesis: A Genetic Approach to Aging*, Medical and Technical Publishing, Lancaster, England, 1974.
5. Britten, R. J.: DNA sequence organization and repeated sequences. *Chromosomes Today* **7**:9–23, 1981.
6. Matsunaga, E.: Perspectives in mutation epidemiology 1. Incidence and prevalence of genetic disease (excluding chromosomal aberrations) in human populations. *Mutat. Res.* **99**:95–128, 1982.
7. Committee on Mutagenicity of Chemicals in Foods, Consumer Products, and the Environment: *Guidelines for the Testing of Chemicals for Mutagencity*, Department of Health and Social Security, Great Britain, 1979.
8. Brewen, J. G., and Preston, R. J.: Analysis of chromosome aberrations in mammalian germ cells. In *Chemical Mutagens: Principles and Methods for Their Detection*, Vol. 5 (A. Hollaender and F. J. de Serres, eds.), Plenum Press, New York, pp. 127–150, 1978.
9. Lejeune, J.: The 21 trisomy—Current stage of chromsomal research. *Prog. Med. Genet.* **3**:144–177, 1964.
10. Bignami, M., Morpurgo, G., Pagliani, R., Carere, A., Conti, G., and DiGiuseppe, G.: Non-disjunction and crossing-over induced by pharmaceutical drugs in *Aspergillus nidulans*. *Mutat. Res.* **26**:159–170, 1974.
11. Knudsen, A. G., Jr.: Mutation and human cancer. *Adv. Cancer Res.* **17**:317–352, 1973.
12. Kalter, H.: Correlation between teratogenic and mutagenic effects of chemicals in mammals. In *Chemical Mutagens: Principles and Methods for Their Detection*, Vol. 6 (A. Hollaender, ed.), Plenum Press, New York, pp. 57–82, 1977.
13. Cacheiro, N. L. A., Russell, L. B., and Swartout, M. S.: Translocations, the predominant cause of total sterility in sons of mice treated with mutagens. *Genetics* **75**:73–91, 1974.

14. Benditt, E. P.: The origins of atherosclerosis. *Sci. Am.* **235**:74, 1977.
15. Guntakatta, M., Matthews, E. J., and Rundell, J. O.: Development of a mouse embryo limb bud cell culture system for the estimation of chemical teratogenic potential. *Teratogen. Carcinogen. Mutagen.* **4**:349–364, 1984.
16. Bauer, K. H.: *Mutationstheorie der Geschwulst-Entstehung. Übergang von Körperzellen in Geschwulstzellen durch Gen-Änderung*, Springer, Berlin, 1928.
17. Burdette, W. J.: The significance of mutation in relation to the origin of tumors: A review. *Cancer Res.* **15**:201, 1955.
18. Miller, J. A., and Miller, E. C.: A survey of molecular aspects of chemical carcinogenesis. *Lab. Invest.* **15**:217, 1966.
19. Ames, B. N., Durston, W. E., Yamasaki, E., and Lee, F. D.: Carcinogens are mutagens: A simple test system combining liver homogenates for activation and bacteria for detection. *Proc. Natl. Acad. Sci. USA* **70**:2281, 1973.
20. Nagao, M., Sugimura, T., and Matsushima, T.: Environmental mutagens and carcinogens. *Ann. Rev. Genet.***12**:117–59, 1978.
21. Brusick, D. J.: The role of short-term testing in carcinogen detection. *Chemosphere* **5**:403–417, 1978.
22. Magee, P. N.: The relationship between mutagenesis, carcinogenesis and teratogenesis. In *Progress in Genetic Toxicology*, Vol. 2 (D. Scott, B. A. Bridges, and F. H. Sobels, eds.), Elsevier/North-Holland, Amsterdam, pp. 15–27, 1977.
23. Setlow, R. B.: Repair-deficient human disorders and cancer. *Nature* **271**:713–717, 1978.
24. Hart, R. W., Setlow, R. B., and Woodhead, A. D.: Evidence that pyrimidine dimers in DNA can give rise to tumors. *Proc. Natl. Acad. Sci. USA* **74**:5574–5578, 1977.
25. Paterson, M. C.: Environmental carcinogenesis and imperfect repair of damaged DNA in *Homo sapiens*: Causal relation revealed by rare hereditary disorders. In *Carcinogens: Identification and Mechanisms of Action* (A. C. Griffin and C. R. Shaw, eds.), Raven Press, New York, pp. 251–276, 1979.
26. Brusick, D.: Mutagenicity and carcinogenicity correlations between bacteria and rodents. In *Cellular Systems for Toxicity Testing 407* (G. M. Williams, V. C. Dunkel, and V. A. Ray, eds.), New York Academy of Sciences, New York, pp. 164–176, 1983.
27. Ames, B. N., and McCann, J.: Validation of the *Salmonella* test: A reply to Rinkus and Legator. *Cancer Res.* **41**:4192–4203, 1981.
28. McCann, J., Choi, E., Yamasaki, E., and Ames, B. N.: Detection of carcinogens as mutagens in the *Salmonella*/microsome test: Assay of 300 chemicals. *Proc. Natl. Acad. Sci. USA* **72**:5135–5139, 1975.
29. Sugimura, T., Sato, S., Nagao, M., Yahagi, T., Matsushima, T., Seino, Y., Takeuchi, M., and Kawachi, T. Overlapping of carcinogens and mutagens. In *Fundamentals in Cancer Prevention* (P. N. Magee, S. Takayoma, T. Sugimura, and T. Matsushima, eds.), University Park Press, Baltimore, pp. 191–215, 1976.
30. Purchase, I. F., Longstaff, E., Ashby, J., Styles, J. A., Anderson, D., Lefevre, P. A., and Westwood, F. R.: An evaluation of 6 short-term tests for detecting organic chemical carcinogens. *Br. J. Cancer* **37**:873–903, 1978.
31. Simmon, V. F. *In vitro* mutagenicity assays of chemical carcinogens and related compounds with *Salmonella typhimurium. Natl. Cancer Inst.* **62**:893–899, 1979.
32. Anderson, M., Binderup, M. L., Kiel, P., and Larsen, H.: Evaluation of the Ames test. *Ugeskr. Laeg.* **14**:305–307, 1979.
33. Rinkus, S. J., and Legator, M. S.: Chemical characterization of 465 known or suspected carcinogens and their correlation with mutagenic activity in the *Salmonella typhimurium* system. *Cancer Res.* **39**:3289–3318, 1979.
34. Bartsch, H., Malaveille, C., Camus, A. M., Martel-Planche, G., Brun, G., Huntefeuille, A., Sabadie, N., Barbin, A., Kuroki, T., Drevon, C., Piccoli, C., and Montesano, R.:

Validation and comparative studies on 180 chemicals with *S. typhimurium* strains and V79 Chinese hamster cells in the presence of various metabolizing systems. *Mutat. Res.* **76**:1–50, 1980.
35. Bridges, B. A., MacGregor, D., and Zeiger, E.: Summary report on the performance of bacterial mutation assays. In *Evaluation of Short-Term Tests for Carcinogens* (F. J. de Serres and J. Ashby, eds.) Elsevier/North-Holland, New York, pp. 49–67, 1981.
36. Zeiger, E.: An overview of genetic toxicity testing in the National Toxicology Program. In *Cellular Systems for Toxicity Testing 407* (G. M. Williams, V. C. Dunkel, and V. A. Ray, eds.), New York Academy of Sciences, New York, pp. 387–394, 1983.
37. Bartsch, H., Tomatis, L., and Malaveille, C.: Mutagenicity and carcinogenicity of environmental chemicals. *Regul. Toxicol. Pharmacol.* **2**:94–105, 1982.
38. Brusick, D. J.: Evaluation of chronic rodent bioassays and Ames assay tests as accurate models for predicting human carcinogens. In *Application of Biological Markers to Carcinogen Testing* (H. A. Milman and S. Sell, eds.), Plenum Press, New York, pp. 153–163, 1982.
39. Auletta, A.: EPA GENETOX Coordinating Committee, 1984.
40. DiCarlo, F. J., and Fung, V. A.: Summary of carcinogenicity data generated by the National Cancer Institute/National Toxicology Program. *Drug Metab. Rev.* **15**(5&6):1251–1273, 1984.
41. Evaluation of the Carcinogenic Risk of Chemicals to Humans. *IARC* October, 1982; suppl 4.
42. Garrett, N. E., Stack, H. F., Gross, M. R., and Waters, M. D.: An analysis of the spectra of genetic activity produced by known or suspected human carcinogens. *Mutat. Res.* **134**:89–111, 1984.
43. Reddy J.: Peroxisome proliferations as nonmutagenic hepatocarcinogens. In *The Toxicology Forum 1985 Annual Winter Meeting Proceedings*, Washington, D.C., pp. 26–40, 1985.
44. Ames, B. N.: Dietary carcinogens and anticarcinogens. *Clin. Toxicol.* **22**:291–301, 1984.
45. Fukushima, S., Shibata, M., Shirai, T., Tamano, S., and Ito, N.: Roles of urinary sodium ion concentration and pH in promotion by ascorbic acid of urinary bladder carcinogenesis in rats. *Cancer Res.* **46**:1623–1626, 1985.
46. Ehling, U. H., Averbeck, D., Cerutti, P. A., Friedman, J., Greim, H., Kolbye, Jr., A. C., and Mendelsohn, M. L.: Review of the evidence for the presence or absence of thresholds in the induction of genetic effects by gentoxic chemicals. *Mutat. Res.* **123**:281–341, 1983.
47. Meselson, M., and Russell, K.: Comparisons of carcinogenic and mutagenic potency. In *Origins of Human Cancer, Book C, Human Risk Assessment* (H. H. Hiatt, J. D. Watson, and J. A. Winstein, eds.), Cold Spring Harbor Laboratory, Cold Spring Harbor, New York, pp. 1473–1481, 1977.
48. Clive, D.: A linear relationship between tumorigenic potency *in vivo* and mutagenic potency at the heterozygous thymidine kinase (TK+/−) locus of L5178Y mouse lymphoma cells coupled with mammalian metabolism. In *Progress in Genetic Toxicology* (D. Scott, B. A. Bridges, and F. H. Sobels, eds.), Elsevier/North-Holland, Amsterdam, pp. 241–247, 1977.
49. Clive, D., Johnson, K. O., Spector, J. F. S., Batson, A. G., and Brown, M. M. M.: Validation and characterization of the L5178Y/TK+/− mouse lymphoma mutagen assay system. *Mutat. Res.* **59**:61–108, 1979.

CHAPTER 4

# Screening Chemicals for Genotoxic Properties

## INTRODUCTION AND BACKGROUND

Much of the data derived from genetic toxicology testing is used in decision-making situations for compound development, although in addition data are increasingly required for regulatory compliance needs.[1] The suitability of a chemical for release into the environment depends on its toxicity, environmental fate, and intended use. The degree of testing must then be combined with regulatory considerations, development and production costs, timing, marketing parameters, and other factors affecting the final decision.[2]

If the chemicals are already in commerce, data defining and quantifying genetic risks following exposure may be needed. Thus, it is difficult to generalize about the choice of methods for chemical testing.

The two major philosophies of testing have been hierarchical (tier) approaches and matrix (battery of tests) schemes. Tier approaches have been most often applied when testing large numbers of substances. In the tier concept, the first level of testing, where all substances must be examined, the tests should be rapid, reliable, and inexpensive.[3,4] At the first level materials can be eliminated on the basis of their responses, and a reduced number of substances will be left to evaluate in the more costly and time-consuming "definitive" tests. The tier concept is a cost-effective program, but it sacrifices time. The reliability of the assays used in the first level of the tier grouping is critical to the success of this approach. If these tests are not reliable screens for genotoxic agents, the tier approach will ultimately be ineffective. Substances passed on to the second

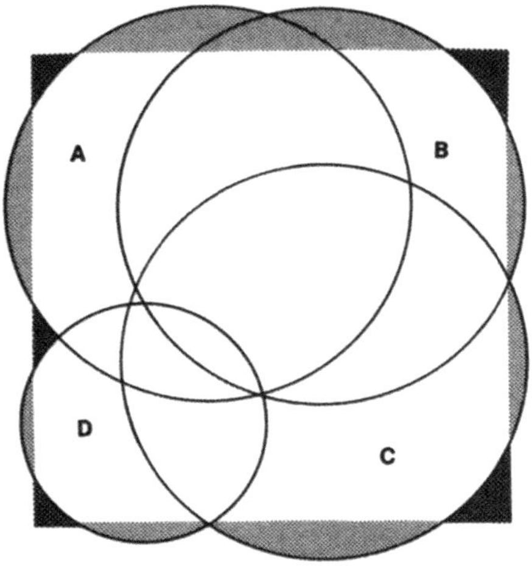


FIGURE 4.1. Battery concept. Each short-term test has its own inherent strengths and limitations, which, when the tests are used in concert, will overlap and compensate for intrinsic limitations. The square represents the testing "window" through which chemicals with genotoxic activity must pass. Each circle represents a specific test and its range of genotoxicity detection. Overlapping indicates compounds detected in multiple assays. Shaded areas represent regions of false responses. Darkly shaded regions represent areas of the "window" that lack adequate test methods. The areas of the circles wil vary according to the tests selected.

and third tiers are subjected to more testing with *in vitro* mammalian cell assays, *Drosophila*, or *in vivo* mammalian models. The intent of the evaluation at the highest level is to develop a qualitative and quantitative estimate of human risk.

The emphasis of testing programs has shifted from a tier approach to a test battery approach. There are several reasons for the shift. Timing of data collection, concern over the possibilities of idiosyncratic responses from primary screening tests, and the need to measure several different genotoxic endpoints even at the first level have all contributed to the growth of the battery philosophy. Mutagenicity testing suggested by EPA, OECD, and other agencies emphasizes this approach by recommending, in some instances, minimal test sets. A battery of tests, including several levels of evaluation conducted simultaneously, provides a window through which only nongenotoxic agents should pass. Figure 4.1 illustrates the rationale for the battery approach to testing. Each test in the battery has its own range of detection (identified as the area of the circle) across a wide range of chemicals. As shown in the figure, some chemicals will fall within the detection area of most or all tests (area of complete overlap of all circles), whereas others will be detected in only one or two tests of the battery. Positive identification of an agent as genotoxic need not be based on positive responses from all the tests in the battery, although replicated positives may reflect the relative concern for hazardous effects from the agent. On the other hand, a chemical identified as nongenotoxic should be inactive in all tests applied. Uncertainty associated with the interpretation of mixed results from test batteries has proven to be their most serious limitation. Because of technical limitations associated with the current state of the art, a few genotoxic agents will be missed by even the best test battery, but careful selection and application of tests will keep this to a minimum.

## DEFINITIONS OF TERMS

Test batteries may consist of screening tests, risk assessment tests, or a mixture of both. It is important at the outset of testing to define terminology carefully and to clarify the relevant biological questions. A committee of the International Commission for Protection against Environmental Mutagens and Carcinogens (ICPEMC) has developed an operational definition for a genotoxic material as "an agent that produces a positive response (e.g., mutation, UDS, chromosome breakage) in any bioassay measuring any genetic end point."[5] While this definition considers virtually all forms of damage of DNA an expression of genotoxicity, the authors stated that "categorization of a chemical as genotoxic does not prove health hazard or risk." This definition is, in fact, only a convenient method of dividing all chemicals into genotoxic or nongenotoxic groups. Readers were advised against presuming that chemicals operationally defined as genotoxic would be hazardous to humans. Additional evidence beyond this initial classification would be required to address concerns of increased genetic risk to somatic or germ cells.

The ICPEMC report also attempted to develop an operational definition for nongenotoxicity. A stringent set of criteria were required to define a nongenotoxin and specified that a chemical be inactive in a series of tests covering five types of endpoints considered critical to the production of all genetic damage. Inactivity in tests measuring the following types of genotoxicity is required to consider a chemical nongenotoxic:

1. Test for alteration of DNA base sequence (nucleotide substitutions, additions, deletions).
2. Test for alteration of DNA integrity (adducts/breaks/cross-links).
3. Test for DNA rearrangement (recombinogenic effects, including SCE).
4. Test for altered chromosome segregation (numerical).
5. Test for altered chromosome integrity (structural).

Table 4.1 illustrates the types of tests that might be assigned to the five types of DNA damage described above. These operational definitions were presented with the sole intent of developing a standardized classi-

TABLE 4.1
Categories of Genotoxicity[a]

| Assay | Alteration of DNA base sequence | Alteration of DNA integrity | Induction of DNA exchange or rearrangement | Alteration of chromosome segregation[b] | Alteration of chromosome integrity |
|---|---|---|---|---|---|
| Dominant-lethal assay | | | | 2° | 1° |
| Mouse specific-locus assay | 1° | | | | 2° |
| Heritable-translocation assay | | | | 2° | 1° |
| Cytogenetics, *in vivo* | | | | | 1° |
| Cytogenetics, *in vitro* | | | | | 1° |
| Sister chromatid exchange (SCE) | | | 1° | | 2° |
| Gene mutation: | | | | | |
| Mammalian cells | 1° | | | | |
| *Drosophila* SLRL | 1° | | | | |
| *Escherichia coli* WP2 | 1° | | | | |
| Salmonella | 1° | | | | |
| Yeast recombination | 2° | | 1° | | |
| Rodent micronucleus | | | | 2° | 1° |
| Bacterial DNA repair | | 1° | | | |
| Mouse spot test | + | | | | + |
| Mouse sperm morphology | Undefined[c] | | | | |

[a] Each test is defined by its primary or established mechanisms, although one cannot rule out that other mechanisms may be involved in producing the observed phenotype or response. 1°, Primary; 2°, secondary, +, equivalent. A minimal test battery at the present time should have at least one test showing a 1°, 2°, or + in each of the categories of genotoxicity. A compound must be negative in all five categories to be considered nongenotoxic.

[b] Several submammalian and mammalian systems are available to test specifically for numerical chromosome anomalies (nondisjunction), but that none has been sufficiently well validated to be recommended for this category of genotoxicity.

[c] A recent publication has reported that altered spermhead morphology can be induced by nongenetic toxicity.[6]

fication scheme for chemicals and not as a method of hazard or risk assessment.

## Screening Tests

Screening tests are those assays that measure the ability of the test substance to produce genotoxic effects. Results from screening tests can define the genotoxic range of an agent but cannot, alone, be used in quantitative risk estimation. In screening tests, attempts are not generally made to relate the applied dose, route of administration, or type of metabolism to the human experience, and allegations of probable health hazards are very difficult to justify if based solely on *in vitro* or submammalian test results. Attempts to extrapolate directly *in vitro* responses to *in vivo* settings have been made, but proponents of this application of screening tests have created confusion and even stimulated resistance to the acceptance of genetic toxicology as a credible source of chemical safety information.

In special cases there may be justification for the development of a risk assessment using results from combined submammalian, *in vitro* mammalian, and *in vivo* somatic tests. For example:

1. Current state of the art in genetic testing with mammalian models limits the assessment of point mutation induction in mammals, especially in mammalian germ cells. While the mouse specific-locus assay may be capable of detecting gene mutations, it is not generally available. Consequently, the most practical of the *in vivo* models rely upon measuring chromosome alterations.[7] A combination of submammalian, *in vitro*, and *in vivo* assays is necessary, therefore, to build a comprehensive data base (including gene mutation) for an evaluation of the genotoxic properties for a particular agent.

2. The target molecule of all genotoxic effects, DNA, is structurally and functionally similar in all organisms. If the test agent is capable of altering the structure or function of DNA in a submammalian species, it can also be expected to produce the same response in rodent and human DNA, providing that the biologically active form of the test agent reaches the appropriate target sites in each case. Thus, submammalian and *in vitro* mutagenicity data coupled with *in vivo* chemical disposition data and germ cell localization studies may, under certain circumstances, be an acceptable alternative to estimate qualitative genetic hazard.[8]

### Hazard Assessment Tests

Tests that exceed the general characteristics of screening tests, yet are not adequate for quantitative risk assessment, are often classified as tests indicating possible human hazard. For the most part these tests consist of somatic cell *in vivo* tests. Bone marrow cytogenetic analysis in rodents and the somatic cell mutation "spot test" in mice are two good examples of tests that are more than screening tests, but are not used in genetic risk analysis.

### Risk Analysis Tests

Genetic risk analysis involves the development of quantitative estimates of transmissible mutation. Tests that qualify for this type of assessment usually measure direct damage to the germ cells or measure heritable effects in the progeny of treated individuals.

A more detailed discussion of tests amenable to quantitative risk assessment is given in Chapter 5.

## CHARACTERISTICS OF ADEQUATE SCREENING TEST METHODS

Because the battery approach to testing now appears to have almost universal acceptance and application, characteristics of tests suitable for a battery need to be described. The number and types of tests available for inclusion in a battery are extensive and include target organisms of divergent phylogenetic levels. At the present time there is no proven method to exclude the use of specific tests; however, "screening" implies a preliminary assessment, which appears to be an appropriate role for the bulk of submammalian and *in vitro* tests.[9]

The following list provides several features of a bioassay that make it a candidate as a screening test.

1. The test should identity agents with specific affinity for DNA, and not agents with nonspecific toxicity for macromolecules. DNA interactions can be measured as primary DNA effects, point (gene) mutations, or chromosome alterations (see Table 4.1).
2. The test should have an adequate capacity for metabolic biotransformation.
3. The test should be reproducible. Study designs should be prepared to ensure adequate sample sizes for definitive, reproducible determination of either positive or negative responses.
4. Test transferability among laboratories should be documented. Tests capable of being performed only in a single laboratory may not be suitable for routine screening.

A 1985 publication derived from the EPA GeneTox program attempts to define the suitability of tests for batteries on the basis of several criteria, including use patterns, laboratory transferability, and extent of protocol standardization.[10] The criteria are listed in Table 4.2 and a listing of the tests identified as "routine" assays is given in Table 4.3. Categorization as routine is not necessarily an indication of the value of a method in screening or risk analysis, and many potentially valuable tests might not be routine methods. Table 4.3 is intended to guide investigators to specific assays about which considerable information exists. This could be of great value in making decisions between two tests that measure the same end-point. Unless other considerations influence the decision, the routine tests should be selected.

TABLE 4.2
Definition of Assay Development[a]

| Level | Required cutoff values: Number of labs | Number of chemicals and classes | Uniformity | Number of tests per year | Definitions |
|---|---|---|---|---|---|
| Routine | >5 | >150 chemicals >10 classes | No significant deviations | >50 | Meets or exceeds all criteria at the levels established |
| Transitional | — | — | — | — | Exceeds all criteria for an "experimental" assay but fails to meet or exceed *all* of those for a "routine" assay (even if it exceeds some criteria) |
| Experimental | 1 or none | <50 <4 | Significant variations among laboratories | <10 | Fails to exceed all criteria at the levels established |

[a] From Brusick and Auletta.[10]

## The Type(s) and Number(s) of Endpoints Detected

Chemicals alter DNA in both specific and nonspecific ways. Point mutations and chromosome rearrangements are specific classes of genetic endpoints. Chemicals also change DNA in such a way that neither point mutations nor chromosome aberrations are specifically detected. Table 4.1 attempts to define the range of genotoxic endpoints that should be included in a comprehensive battery to establish a nongenotoxin. *In vitro* cell transformation studies are often included in test batteries, but are not identified in Table 4.1, because the precise nature of the molecular change in unknown. The evidence for cell transformation *in vitro* is generally a mass (focus or clone) of cells with altered morphology, growth patterns, and biochemistry.[11] Transformed cells are capable of proliferating into a tumor when implanted into the appropriate host animal[12,13] and it is possible that *in vitro* transformation assays may mimic some of the molecular events that occur during oncogenesis *in vivo*.

## The Metabolic Capability of the Test or Associated Activating System

The adaptation of host-mediated and S9 microsome activation systems to bioassays has enhanced the detecting capacity of most *in vitro*

TABLE 4.3
Summary of the Characteristics of the Group of Tests Identified as Routine by the EPA Gene-Tox Program[a]

| Test name | Test type | | | Type of lesion detected | | | Applicable to germ cell risk |
|---|---|---|---|---|---|---|---|
| | Phylogenetic status | *In vitro* | *In vivo* | Gene mutation | Chromosome aberration | Direct DNA damage | |
| *Salmonella typhimurium*/microsome reverse mutation | Bacteria | X | | X | | | No |
| *Escherchia coli* reverse mutation | Bacteria | X | | X | | | No |
| *Escherchia coli* DNA repair test | Bacteria | X | | | | X | No |
| *Drosophila* SLRL test | Insect | | X | X | | | Yes |
| Chromosome analysis | Mammalian cells | X | | | X | | No |
| SCE (all sister chromatid exchange tests) | Mammalian cells | X | | | | X | No |
| V79 HGPRT mutation analysis | Mammalian cells | X | | X | | | No |
| Bone marrow chromosome analysis | Rodent | | X | | X | | No |
| Micronucleus test | Rodent | | X | | X | | No |
| Dominant lethal test | Rodent | | X | | X | | Yes |

[a] From Brusick and Auletta.[10] The use of data employed to identify these tests as routine were derived from Gene-Tox reviews of test results through 1980–1981.

TABLE 4.4
Types of Chemical Activating Systems That Are Used with Screening Test

| Activating system | Reference | Advantages/limitations of the method |
|---|---|---|
| Host-mediated assays (HMA) | 16 | All have the advantage of reflecting the typical *in vivo* activation/detoxification balance; their major problem involves implantation, localization, and recovery of indicator cells within the host |
| Intraperitoneal | 17 | |
| Intrasanguine | 18 | |
| Urine analysis | 19 | |
| Other | | |
| Microsomal enzyme preparations | | The primary advantage is that good contact between chemical and enzymes can be assured in the microsome mix; the response interpretations are complicated by the loss of all compound activation/detoxification interactions as well as by the coordination in conjugating (glutathione) and deconjugating enzyme systems |
| Purified microsomes | 20 | |
| S20 preparations | 21 | |
| S9 preparations | 22 | |
| Crude homogenates | 23 | |
| Cell-mediated systems | | These systems provide good metabolic balance and simulate the metabolic systems as found in the intact liver; they also maintain proximity between test substance and activating enzymes; the cell-mediated systems do not require cofactor reaction mixtures; the limitations rest with the technical difficulties of preparing the cell systems to the test methodology; this approach appears most promising with mammalian cell culture techniques |
| Primary hepatocytes | 24 | |
| Irradiated feeder cells | 25 | |

bacterial tests. Other submammalian organisms, such as *Drosophila* and some of the fungi, have a broad range of intrinsic metabolic activation.[14,15] Mammalian cells used in genetic assays have some endogenous metabolic capability, but except for some primary hepatocyte cultures, they must be coupled with some form of exogenous activation before use in screening. Table 4.4 describes some of the commonly used activation methods.

*In vitro* metabolism is a complex and controversial subject (see Appendix A). *In vitro* microsome-mediated metabolism offers an isolated view of the biological properties of activated forms of the test substance, but often fails to handle the parent compound and its products in the same way as *in vivo* metabolic pathways. This subject is dealt with in many review articles and books.[26,27] A substantial body of test data has been generated using microsomal activation systems, but for many agents microsome systems provide only rough approximation of *in vivo* events. The use of isolated fresh hepatocytes probably gives a better representation of *in vivo* effects than microsome fractions, especially for a balance of activation and detoxification processes. S9 mixes favor activation, since several detoxifying enzymes, such as glutathione transferase, are inoperative.

The following generalizations regarding the routine application of activation systems to supplement endogenous target cell capabilities may be helpful in constructing screening protocols:

1. Animal host-mediated systems are more likely than *in vitro* microsome preparations to generate false-negative results. There are many possible reasons for this situation, but limitations of the accumulation of metabolites and their systemic distribution are the most likely.

2. Interspecies comparisons of microsomal preparations have shown clear quantitative differences in metabolic activation, but very little evidence exists for unequivocal qualitative differences. This includes studies with human S9 preparations. Rodent S9 preparations appear to give profiles of activation equivalent to those obtained with human activation reactions.[28] Species and organ differences are most divergent with uninduced enzymes; most species differences disappear when hepatic inducers such as PCBs are used in S9 preparation. Recent reports on the results from NCI- and NIEHS-supported testing with the Ames assay indicate that S9 prepared from Syrian hamster liver may offer some advantages over the rat for specific chemical classes.[29]

3. Although various methods of microsomal enzyme induction have been reported, the use of Aroclor 1254 produces a broad-spectrum induction and seems to be adequate for detecting most genotoxic substances. Induced microsomes are preferred over noninduced preparations for general screening.

4. There does not appear to be any advantage in using microsome preparations cleaner than S9 (e.g., S20 or purified microsomes), and crude tissue homogenates appear less optimal than S9 preparations.

5. Little benefit has been derived from the use of activating enzymes from tissues other than liver. While quantitative differences can be observed among tissues, the results with liver coincide with activity detected using S9 from other organs. Organ differences could be useful in special studies addressing target organ susceptibility, but do not contribute much to screening programs, because the variation in activation appears more quantitative than qualitative.[30] It may be desirable in some instances to match species when performing *in vitro* activation studies and to compare those data with other toxicity data collected *in vivo*, especially if mutagenic activity is encountered.

## Test Reliability/Reproducibility

There are been considerable concern among scientists concerning the potential problems associated with intra- and interlaboratory reproducibility of biological assay results. This concern has been most vocally expressed for submammalian and *in vitro* mammalian assays. There are a number of reasons for this, some legitimate and some not. It is possible for two laboratories to disagree on the genotoxic potential of a given substance or, in some cases, for the same substance to be declared both active and not active on two different occasions with the same test in a single laboratory. This level of variability exists in all biological systems, especially when results are expressed in binary responses (+ or −) derived from arbitrary cutoff points rather than specific values. Variability also increases as the biological activity (potency) of a genotoxin decreases or disappears. Some of the reasons for response inconsistency are as follows:

1. Inexperienced Investigators. Variable test results often stem from the almost unlimited access to *in vitro* test systems by anyone wishing to conduct an evaluation, including individuals with little or no training in genetics. For example, the *Salmonella* strains used in the Ames assay are in virtually all microbiology laboratories in the United States, and the cost of conducting the test is minimal. Without adherence to standardized test methods or adequate quality control procedures, test anomalies will certainly arise, and if the individuals conducting the test are not thoroughly experienced with the assay, it is quite possible that aberrant results or artifacts will not be identified. Once aberrant results are reported to regulatory agencies or are sanctified by journal publication, it is extremely difficult to negate their impact even if it can later be established that the effects were the result of protocol deficiencies.
2. Variability in Data Analysis. Another cause of apparent conflicting results is the investigator's approach to data evaluation and interpretation. Current methods of evaluation are often intuitive and similar sets of data may be interpreted differently by two investigators. Each investigator's approach to interpretation of the data will be affected by previous experience with the test system and by the evaluation criteria and statistical analyses employed. Possible solutions to this source of variability are presented later in this chapter.
3. Protocol. The size and therefore the power of the test also contribute to inconsistent interpretations. Comparison of data from two trials of an assay, one using minimal sample sizes and one using a large-scale study design having a greater resolving power, may yield different conclusions. It is therefore extremely important in making intra- and inter-

laboratory comparisons that careful analysis of the data be accomplished by thoroughly examining the comparability of study designs, sample sizes, and scoring procedures. Investigator bias via prior knowledge of the structure or toxicological properties of a compound has been suggested as the reason for exceptionally high correlations between expected and obtained responses in some validation trials.[31] Elimination of investigator bias through the use of coded samples, slides, and plates becomes critical in developing useful validation information.

4. Test Response Confirmation. Even when all known confounding factors of studies are controlled, there will be occasional sets of conflicting test results. For this reason, it is recommended, when feasible, to incorporate an automatic, confirmatory retest sequence into chemical screening programs. Current testing guidelines issued by EPA (OPTS) and OECD recommend independent repeats on all *in vitro* and submammalian tests. If all or a portion of each test, regardless of the response, is repeated after a specified interval following the first test and the results match the interpretation, the possibility of test anomalies or experimental error accounting for the results is significantly reduced. If the two trials conflict, review of the procedures followed by additional testing may be necessary to resolve the conflict. Obviously, this is a more costly and time-consuming approach to screening, but reproducible results increase the reliability of the developing data base and facilitate scientific and/or regulatory decisions. The independent trial approach will also benefit the science in that a reproducibility or reliability quotient can be ascertained both for the test and the performing laboratory.

### Test Transferability

This important consideration is often unappreciated by laboratory management and investigators. Many attractive techniques for measuring genotoxicity have been reported, but when different investigators attempt to apply the method to other chemical classes or to reproduce the published studies, they often obtain unexpected responses. Examples of tests that have proven difficult to transfer across labs include alkaline elution techniques for DNA breakage, anaphase cytogenetic analysis, cell transformation in primary cell cultures, and some types of host-mediated assays.

## STRATEGIES FOR BUILDING A TEST BATTERY

### General Philosophy

Test selection can be based on a variety of themes.[9,32] For the most part, a test battery represents the experience and bias of the investigator

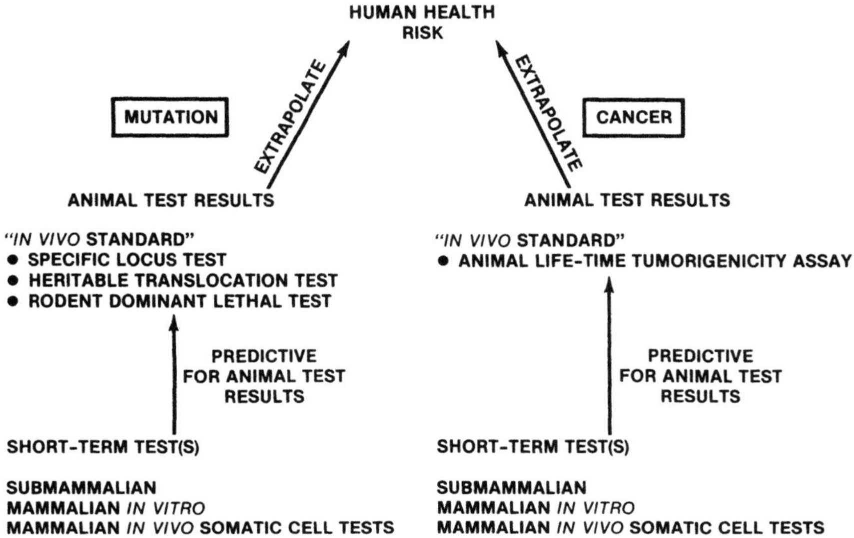


FIGURE 4.2. Description of the approaches followed to select short-term tests for prediction of animal mutagens and carcinogens. The basic assumption is that animal test data are the most appropriate for developing accurate risk assessments extrapolatable to humans. Because of the cost and time involved in performing the *in vivo* tests used as standards, predictive short-term tests can become cost-effective components in routine screening of chemicals. The short-term tests used in both parallel pathways may have considerable overlap, but each group should be determined by selecting those assays that most accurately reflect the responses in the corresponding "*in vivo* standard." The diversity of the chemical spectrum in each *in vivo* standard will exert its own discriminatory influence on the test selection. Tests should be selected based on performance in reflecting the *in vivo* responses plus other practical considerations of test performance (cost, facilities, performance time, data analysis).

as well as the financial resources available to conduct tests. Some batteries are constructed for carcinogenicity prediction; others encompass genetic risk, with emphasis on rodent germ-cell mutagenesis. Accomplishing both objectives (detection of animal mutagens and carcinogens) with a single test battery is unlikely because the most suitable test battery for each type of toxic phenomenon is derived from an independent set of data (Figure 4.2).

## Specific Approaches

1. Database Approach. This approach was first recommended by Ray.[33] The EPA-sponsored Gene-Tox program also has as one of its objectives the development of test batteries.[34] Results of the Working Group

reports are to be evaluated by an EPA steering committee for test performance with the intent of defining special test batteries optimal for detecting genotoxins in specific chemical classes. It is also presumed that analysis of the entire data set would facilitate the emergence of a general test battery for carcinogenicity potential. This process is still underway; however, the activity profile of data base is heavily skewed toward carcinogens and positive test responses, making accurate estimates of assay specificity difficult.

Another approach has been taken by Rosenkranz and colleagues at Case Western Reserve University.[35] Using the Gene-Tox data base and a computer analysis scheme, this group developed the Carcinogen Prediction and Battery Selection (CPBS) system. This software uses the reported reliability indexes of various tests as predictors of animal carcinogenicity to suggest test batteries for unknown agents and predict carcinogenic potential based upon test results.

2. Base Plus Complementary Test Approach. This approach has been suggested primarily for carcinogen detection. The method rests on the use of a well-established, routine assay, with extensive testing experience against carcinogens. This experience will define the strengths and weaknesses of the base test. A complementary test or set of tests known to compensate for the limitations of the base test are then recommended to complete the test battery. This approach has been suggested by Weisberger and Williams[36] and Ashby *et al.*[37] Weisberger and Williams combined the Ames assay and the rat hepatocyte UDS assay, and Ashby recommends the use of the Ames test plus an *in vitro* chromosome aberration assay.

3. Endpoint/Phylogenetic Approach. This approach, unlike the previous two methods, does not rely upon retrospective analysis of test performance in the selection of assays. It rests upon the assumption that a reliable test battery should cover the major mechanisms of genetic toxicity and should contain a variety of organisms with diverse phylogenetic complexity. An example of this approach is the genetic toxicology requirements under the EPA's pesticide registration program. These requirements call for at least three assays, including gene mutation, chromosomal damage, and primary DNA damage. Some batteries also add cell transformation as a separate endpoint. The need for phylogenetic diversity is not explicitly stated, although encouraged.

The Endpoint/Phylogenetic Approach is not tied to either carcinogenic or mutagenic prediction, but rather addresses the broader view of genetic toxicity. Through the selection of individual tests one might structure the approach toward one or the other of those two toxic phenomena.

A final consideration relevant to all of the previously described approaches is how many assays to include in the battery. This topic has been the subject of several analyses, but is not resolved.[38]

A general principle is that batteries with a few tests (e.g., two or three) are more susceptible to false-negative outcomes, whereas a battery with more than five tests is more likely to have a false positive among the responses. Batteries skewed toward a single genetic endpoint or a single type of organism are also susceptible to false responses. Because this approach does not use test performance *per se* in test selection, an investigator may have several test options for each endpoint or phylogenetic level. Assistance in selection of tests is possible through the use of a publication from the EPA Gene-Tox program that attempts to define the maturity or developmental level of an assay of *in vitro* and *in vivo* tests.[10] One can therefore select for each endpoint or phylogenetic level a test that is well established and considered a routine assay (Table 4.3).

## CONDUCTING THE TESTS

The selection and execution of protocols or study designs is as important to the genotoxic assessment of an agent as the selection of tests for the battery. Study designs should meet the Good Laboratory Practices requirements and should be of sufficient size (power) to give good experimental resolution. Standard study designs for most of the routine genetic toxicology assays have been published by EPA (1982; 1985),[39] OECD (1981; 1983),[23] and EEC (1983).[40]

### General Criteria for Dose Selection

Dose selection is important in all biological testing, but it is critical with *in vitro* testing. The use of improper treatment conditions is one of the most common causes of artifactual responses in genetic toxicology testing. Test concentrations should be high enough to demonstrate physiologic effects in the target cell population, yet not so high as to produce secondary cellular effects capable of confounding data interpretation. Cultured mammalian cells, for example, are particularly susceptible to mutation, chromosome disruption, and cell transformation caused by ionic or pH levels that exceed normal physiologic concentrations found *in vivo*.[41]

Most free-living organisms, single-cell or multicellular, have evolved mechanisms to adjust to rapid changes in their environment. For example, a bacterium subjected to rapid changes in pH survives by the regulation of ion transport into and out of the cell. In mammals, pH regulation is mediated not only by ion transport, but also by a complex set of organic

TABLE 4.5
Salt-Induced Mutant Frequencies in L5178Y Mouse Lymphoma Cells[a]

| | Dose | | Test conditions | | | |
|---|---|---|---|---|---|---|
| | | | −S9 | | +S9 | |
| Salt | μg/ml | mM | MF × $10^{-6}$ | RTG (%) | MF × $10^{-6}$ | RTG (%) |
| NaCl | 0 | 0 | 28 | 100 | 58 | 100 |
| | 4000 | 68 | 57 | 40 | 81 | 21 |
| | 5000 | 86 | 134 | 12 | 59 | 17 |
| | 0 | 0 | 40 | 100 | 81 | 100 |
| | 5000 | 86 | 67 | 43 | 123 | 71 |
| | 6000 | 103 | 75 | 35 | — | 44 |
| KCl | 0 | 0 | 26 | 100 | 24 | 100 |
| | 5000 | 67 | 28 | 65 | 43 | 36 |
| | 0 | 0 | 34 | 100 | 44 | 100 |
| | 4000 | 54 | 28 | 69 | 73 | 26 |
| | 5000 | 67 | 40 | 55 | 69 | 17 |
| | 7000 | 94 | 76 | 30 | — | — |
| | 8000 | 107 | 115 | 12 | — | — |
| KF | 0 | 0 | 45 | 100 | 31 | 100 |
| | 400 | 6.9 | 84 | 46 | — | — |
| | 500 | 8.6 | 72 | 43 | 76 | 12 |
| | 600 | 10.3 | 90 | 22 | — | — |
| | 0 | 0 | 45 | 100 | 71 | — |
| | 500 | 8.6 | 72 | 46 | 136 | 18 |
| | 600 | 10.3 | 76 | 22 | — | — |
| | 700 | 12.0 | 118 | 5 | — | — |
| NaF | 0 | 0 | 29 | 100 | 26 | 100 |
| | 400 | 9.5 | 45 | 41 | 39 | 41 |
| | 500 | 11.9 | 55 | 17 | 81 | 13 |
| | 600 | 14.3 | 83 | 10 | 94 | 8 |
| | 0 | 0 | 25 | 100 | 33 | 100 |
| | 500 | 11.9 | 41 | 16 | 76 | 21 |

[a] Average mutant frequencies (MF) for triplicate treatments and quadruplicate controls are given; other treatments below the positive ranges were deleted. Underlined values are considered positive responses. RTG, Relative total growth. Test medium, Fischer's medium plus 5% horse serum. (Data courtesy of B. Myhr.)

buffers in serum and other tissue fluids. Explanted cells cultured *in vitro* lack the capacity for environmental adjustment that exists in free-living single-cell organisms.

Important phenomena have been observed in cultured mammalian cells exposed to pH and osmotic conditions exceeding normal *in vivo* ranges. Reduction of pH to a range of 5.5 ± 0.5 resulted in genotoxic effects expressed as chromosome breakage and mutation.[42,43] While a pH

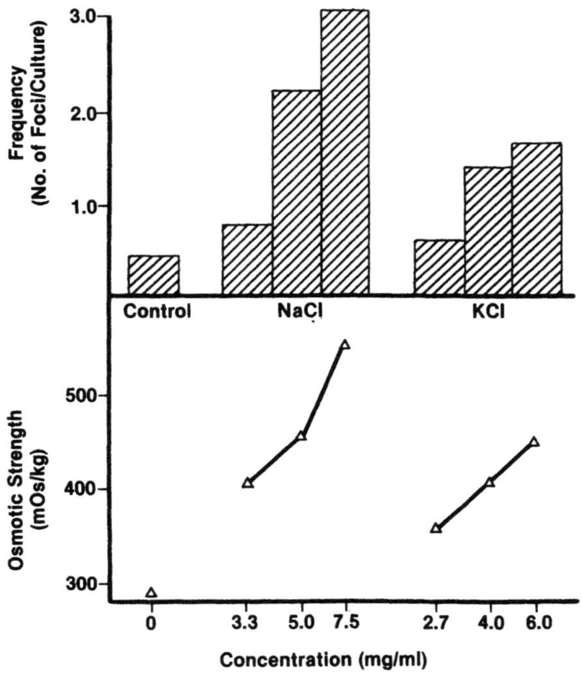


FIGURE 4.3. Transforming activity of NaCl and KCl using BALB/c–3T3 cells. (Data courtesy of J. O. Rundell and E. Matthews.)

of 5.5 does not pose a serious threat to the viability of cells in culture nor for isolated cell populations in mammals (i.e., stomach lining), it would have a serious impact on the viability of the intact animal. The mechanisms responsible for genotoxicity induced by low pH levels are not well understood, but identification of this phenomenon has substantial impact on how dose levels are established for *in vitro* assays. The activity of low pH as a mutagen in bacteriophage and the modulation of chemical mutagenicity in bacteria and yeast by low pH have been previously reported.[44]

Changes in the osmotic levels of tissue culture media also affect the genetic integrity of cultured mammalian cells. Several ionic species, such as sodium and potassium, result in chromosome damage, gene mutation, and cell transformation at environmental levels raising the media osmolality above 400 mOs/kg.[41,45] This phenomenon was recognized by Ishidate *et al.*[46] in conducting chromosome aberration studies. Results of mutation analysis of sodium and potassium ions in the mouse lymphoma assay suggested that different ion species have differing mutagenic activity

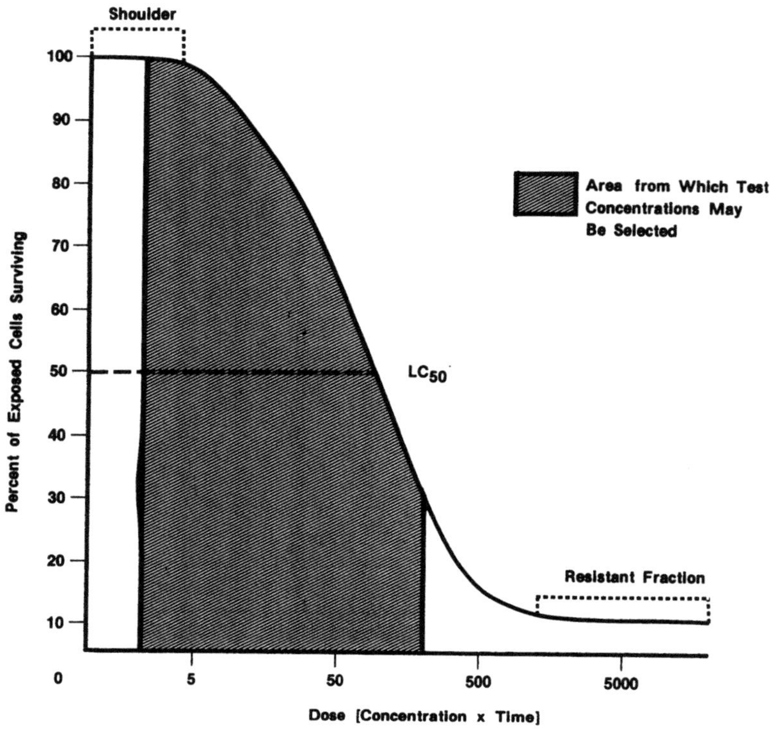


FIGURE 4.4. Typical survival curve for target cells used in screening-type assays. Preliminary toxicity studies generate survival curves and are useful in selecting test concentrations. The shoulder probably represents the area of effective repair of all lesions. At least one concentration should fall into this region of nontoxicity. The $LC_{50}$ is the lethal concentration for approximately 50% of the target cells. The resistant fraction may represent a unique subpopulation and the concentration range should not include this region.

(Table 4.5) and that both the cation and anion may influence the mutagenic response (e.g., NaFl is a stronger mutagen than NaCl).[41] Matthews and Rundell also reported (Figure 4.3) the transforming activity of NaCl and KCl using BALB/c–3T3 cells.[41]

Ashby[41] suggested a mechanism through which high levels of monovalent ions lead to replacement of the divalent magnesium ion in chromatin, rendering it susceptible to breakage. This effect seems to be threshold-dependent, occurring at 400 mOs/kg and above.

Protocols for screening unknown chemicals usually outline steps for dose range selection, including a definition of the cytotoxicity curve (see Figure 4.4). Some chemicals, however, are virtually nontoxic, and selection of maximum testing concentrations of 5–10 mg/ml for these agents

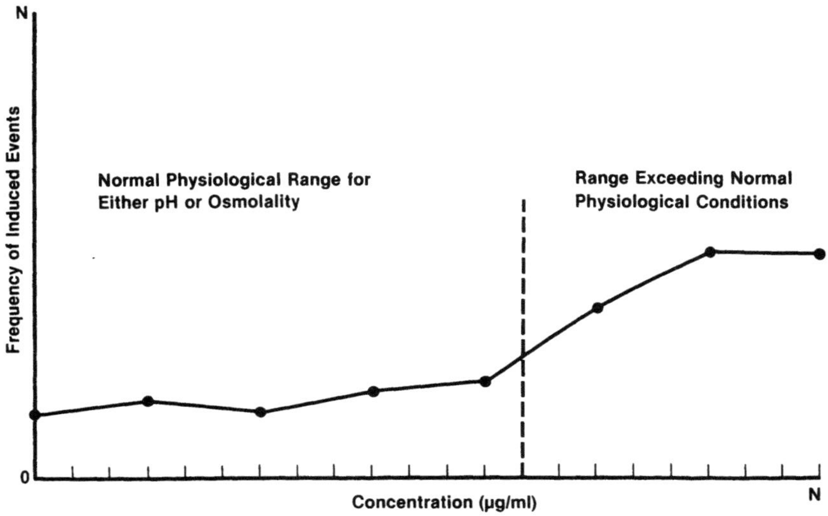


FIGURE 4.5. Genotoxicity results obtained from a region where the normal homeostatic ranges for pH or osmolality are exceeded might be the result of treatment conditions and not the test chemical. Positive responses in these ranges should be viewed with caution and modifications in testing should be considered. (-●-) Effect measured (e.g., aberrations, mutation). The dashed vertical line separates the regions of normal and nonphysiological treatment conditions.

is common. Positive response found at extremely high ion levels or significantly altered pH might not extrapolate to *in vivo* responses. Treatment conditions can influence the outcome of cell culture assays and should be carefully controlled to normal homeostatic levels to ensure interpretable data. Maximum testing concentrations selected on the basis of cell killing may be inappropriate for some chemical types.

It is not possible to fix a standard dose range for each assay, but Figure 4.4 illustrates the area on the toxicity curve from which test concentrations for *in vitro* assays are generally selected. The number of points selected to cover this range will depend on the nature of the test and the desired resolution. Generally, the more points sampled, the more reliable the results. Some genotoxic agents may have a narrow range of activity, which coincides with solubility, pH, or other factors. Testing of compounds that do not show a toxicity curve similar to that in Figure 4.4 but are essentially nontoxic over a range of concentrations exceeding 5 mg/ml or the limits of solubility should have dose selection influenced by factors of pH and/or osmolality. If concentrations exceeding normal homeostatic ranges for pH and osmolality are used, interpretation of positive test results should take these deviations into consideration (Figure 4.5).

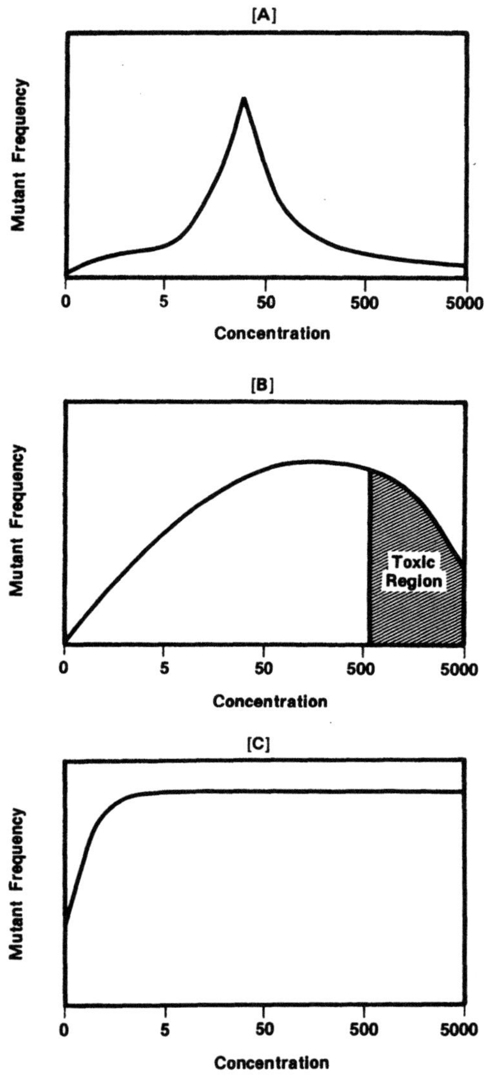


FIGURE 4.6. Types of dose–response curves obtained in screening tests. Three relatively typical types of dose–response curves are shown. The variations in shape emphasize the need to employ multiple-dose protocols for chemical screening. (A) Very narrow range, which may be missed if concentration steps are too far apart. Not very typical, but can occur if pH range or osmotic conditions are critical. The compound is basically nontoxic. (B) Most typical type of response curve. One may get a set of data with an inverse concentration response if dose range is selected too high. (C) Some chemicals have only a short range over which a dose response is observed. A plateau is reached because of solubility, cell permeability restrictions, or other rate-limiting processes. The compound is basically nontoxic.

Figure 4.6 shows some examples of hypothetical dose–response cuves, and identifies some of the problems encountered in dose range development. For example, the type of response generated in Figure 4.6A might easily be missed if the doses are spread too widely. Figure 4.6C the dose–response portion of the curve is missed even at the lowest concentrations, making analysis difficult, since all concentrations fall on the plateau. The presence of several doses on the downside portion of the response curve in Figures 4.6A and 4.6B also creates interpretation problems. These responses appear as an inverse dose relationship. These examples illustrate the importance of careful selection of concentrations toward the lower end of the concentration range. General recommendations for selection of *in vivo* dose levels are given in Appendix B.

Cytotoxicity and dose selection are involved in another problem area. Scoring for some genetic events often requires a selective medium that will not support the growth of nonmutant cells, thus requiring a separate medium to determine surviving populations. This occurs most commonly in mutation studies where cell survival is derived from counting colony-forming units (CFUs) from dilutions of the treated population under nonselective conditions. There is the possibility of selection, that is, that the test material preferentially kills the nonmutant cells at a different rate than mutant cells, resulting in an artifact that resembles mutation induction. Performing lethality tests on separate populations of normal and mutant cells with the test substance may identify cell-type specificity. If selection is occurring, the nonmutant cells will show greater sensitivity (lower $LC_{50}$). Another indication of selection is an absence of absolute increase in genetic events (mutants, aberrations, etc.) over the dose range employed. If, irrespective of survival, scored events increase with dose, then selection may tentatively be ruled out. Figure 4.7 shows what may happen to mutant frequencies if they are calculated under conditions of extreme toxicity and low cell survival. The results shown in the figure may not represent true mutation induction.

## The Use of Controls in *in Vitro* and *in Vivo* Testing

Many, but not all, toxicology studies include concurrent controls. The most generally applied controls are those for the solvent or vehicle used in the preparation and administration of the test substance. In addition to the solvent/vehicle control, other negative controls (assay system minus solvent/vehicle and test substance) are occasionally conducted to determine the true spontaneous background level of the assay system. Depending on the type of assay and the inherent variability, it is also possible to compare the test results with a composite historical control data base. This comparison can be valuable for certain types of studies, especially when small sample sizes are involved. Positive controls are the least consistently applied in toxicology testing. A positive control consists of performing a study similar to that used for the test sample with an agent expected to produce an effect in the assay system. There are several purposes for conducting positive controls. Table 4.6 summarizes the rationale for commonly employed controls.

Negative and solvent/vehicle controls have obvious importance in

| Concentration [μg/ml] | Surviving Cell Fraction [x 10^4] | Actual Count of Mutants Recovered | Calculated Frequency [x 10^-6] |
|---|---|---|---|
| 0 | 312 | 23 | 7.37 |
| 1 | 287 | 21 | 7.31 |
| 3 | 190 | 20 | 10.52 |
| 10 | 113 | 19 | 16.81 |
| 33 | 58 | 17 | 29.31 |
| 100 | 40 | 18 | 45.00 |
| 333 | 26 | 13 | 50.00 |
| 1,000 | 7 | 3 | 42.85 |

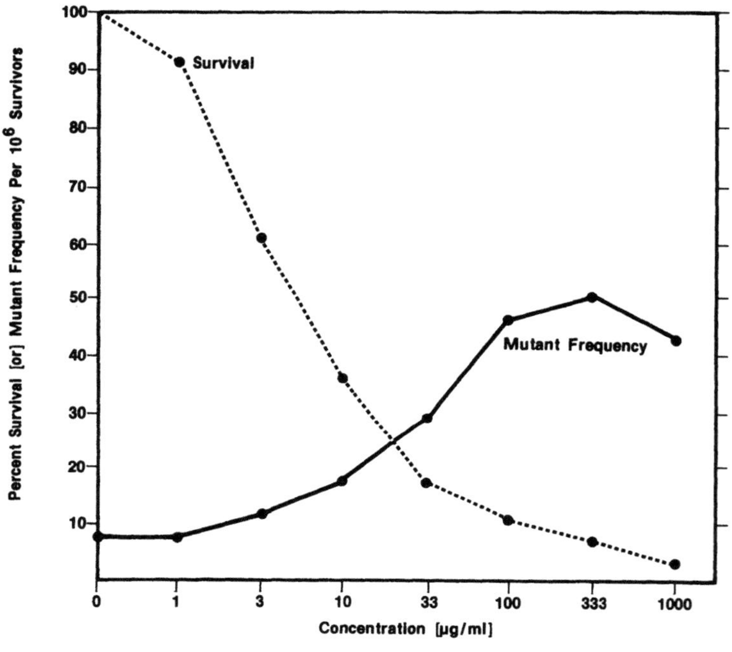


FIGURE 4.7. Potential selection error in mutation dose–response study using a hypothetical situation. It is possible to produce dose–response mutation induction curves through improper dose selection. The mutant frequency shown appears to increase with dose, but close inspection of the test data shows that there was no net increase in the number of mutants. The frequencies increased because of an apparent faster rate of killing for the nonmutant cells. These data raises the possibility of selection rather than mutation. Careful evaluation of data is required in cases of this nature.

conducting toxicology tests. These data provide the no-effect or baseline response from which all statistical or other types of data interpretation derive. Reliable data analysis would be impossible without the inclusion of these controls in a study. Occasionally, however, studies require such enormous populations of organisms to demonstrate a background fre-

TABLE 4.6
Common Types of Control Studies Employed in Toxicology

| Untreated negative control | Solvent/vehicle control | Historical negative control | Positive control |
|---|---|---|---|
| Consists of the assay system without test agent or solvent/vehicle added | Tests of the solvent or vehicle used in the preparation and administration of the test substance; if a solution is obtained, the control is designated solvent control; if the test material is not solubilized, the control is referred to as a vehicle control | A composite of previous control data collected from similar type studies | Consists in conducting the assay with a substance known to produce an effect in the test |
| Defines the inherent background level of effects in the test | Provides the baseline to which the test data are compared for significant effects; any effects of the solvent/vehicle are assumed to be corrected for in this control | This control is a valuable aid in putting data into perspective, especially when the sample sizes for the concurrent control are relatively small; this control also permits variations in background levels due to environmental or temporal cyclic shifts to be compensated for in data analysis | Is not generally used as a baseline for interpretation of test substance data |
| Gives a baseline from which test data and solvent/vehicle data can be interpreted | | Is not essential in a minimal test, but may have critical value in certain types of biological assays | Is not generally considered essential in a minimal test in which an adequate solvent/vehicle control has been included |
| Is not essential in a minimal test since the solvent control provides a baseline for interpretation; in tests that use no solvent/vehicle, however, this is an essential component | | | May be used to assess the ability of the investigator to produce and interpret effects in the test system |
| | | | May be a control for *in vitro* metabolic activation system if activation-dependent chemicals are tested |

quency or rate that the test data collected are routinely compared to an historical or cumulative data base. Certain types of animal studies, such as the heritable translocation assay (HTA) in mice or the sex-linked recessive lethal (SLRL) assay in *Drosophila*, can be evaluated in this manner.

In the case of the mouse specific-locus test and the heritable translocation assay, an historical control is usually involved in the data evaluation, since the likelihood of finding a spontaneous translocation in the typical concurrent control populations is extremely low.[47] This is one of the few instances in which historical control data are more important than concurrent control data. Some laboratories use historical values for bone marrow cytogenetic analysis and conduct only a small concurrent solvent control to check for unusual variation due to disease or other modifying parameters.

Similarly, in *Drosophila* testing for recessive lethal, a cumulative laboratory control value is often used in interpreting data. In this case, the laboratory cumulative control consists of many thousands of individuals and may be critical in data interpretation. Statistical tables giving the requirements for test sample size have been published for *Drosophila* studies using the average historical control frequency as a selection parameter.[48] The frequency of *Drosophila* SLRL in a given laboratory might be 0.20% based on accumulated control data from 50,000 flies. A single experiment is conducted with a small control consisting of 1000 flies and several concentrations of a test substance. Assume that the concurrent control frequency based on 1000 flies is 0.12% and that the recessive lethal frequency at the highest concentration of the test substance is 0.30%. While this frequency is almost threefold higher than the concurrent control, the cumulative control results might be viewed as eliminating the 0.30% value as a positive effect. Without the benefit of the historical data and the experience of what types of variation the controls show with small populations, data may occasionally be incorrectly evaluated.

Valid tests can be conducted without a concurrent positive control test for each test run. In fact, most animal-based toxicologic testing is performed without such controls. Most of the purposes for using positive controls can be evaluated by careful analysis of negative or solvent/vehicle control data. The proper functioning of a test system can be evaluated by examination of the viability, growth rate, and physiologic characteristics of the test organism coupled with the frequency with which the endpoint occurs spontaneously. Similarly, an investigator's expertise can be determined through evaluation of scored spontaneous events.

In genetic toxicology tests involving bacteria, the inclusion of positive control values takes on more significance because it is not practical to examine the test organisms directly; their capacity to respond to known genotoxic agents is the most reliable method for quality control. Frequently, in microbial or mammalian *in vitro* cell culture testing, an activation (S9 mix) system is employed for the biotransformation of xenobiotic molecules to intermediates. The proper functioning of this activation system can be measured with chemicals that produce positive responses following enzymatic biotransformation. Thus, positive controls for activation studies serve the additional function of quality control for the S9 activation system. When substances are evaluated in a test, both with and without an S9 activation system, the activation positive control alone provides all of the needed positive control functions.

Concentration selection for positive control compounds is important. Ideally, the best concentration is high enough to show a consistently significant response, but low enough so that suboptimal treatment conditions will be detected by a loss of activity.

Compound selection for positive controls is also important. Similarity in structure to the test substance is helpful but often not possible, especially in chemical screening of large numbers of diverse agents. Demonstrated activity with a single positive control of one chemical class does not automatically optimize the test system for chemicals of other classes.

## Use of *in Vivo* Genetic Toxicology Assays

Some of the considerations in the use of genetic toxicology evaluations are:

1. What are the appropriate circumstances to introduce *in vivo* tests into a chemical assessment?
2. What *in vivo* tests are available and for what purpose is each best suited?
3. What information will be gained from *in vivo* studies, especially with regard to the interpretation of existing *in vitro* results?

The EPA has proposed guidelines for the interpretation of data, including *in vivo* tests, in the context of a weight-of-evidence approach to genetic risk.[8] These guidelines provide a framework for evaluation, but do not address when or how to select tests. Another comprehensive analysis of genetic risk assessment methods was prepared by ICPEMC Committee 4.[49] This report is a technical analysis of risk analysis methods and is not directly applicable to general testing programs.

Once an *in vitro* test battery has been completed and evaluated it may be necessary to extend the testing to *in vivo* assays. The following set of scenarios may be appropriate to trigger *in vivo* studies:

1. *In vitro* test battery responses are mixed with a key genetic end-

point (e.g., gene mutation or chromosome breakage) registering a clear positive response. The *in vivo* data are important to put this response into perspective.
2. *In vitro* test battery responses are mixed and positive responses are found in more than one genotoxicity assay (e.g., DNA binding, UDS, SCE, cell transformation), but all mutation and chromosome aberration endpoints are negative. *In vivo* negative effects for mutation and clastogenesis would reduce concern for genetic hazard that might have been raised by the other findings.
3. *In vitro* data are uniformly negative, but there is some reason to believe that *in vivo* metabolism is not represented in activation systems used *in vitro*. Negative *in vivo* results will support and provide additional evidence against a potential health risk.
4. *In vitro* responses are all negative, but the extent of compound use and human exposure levels necessitate extension of the data base into those that involve germ cells in addition to somatic cells.

The following other circumstances unrelated to *in vitro* responses can warrant *in vivo* analysis:

1. Cancer bioassay results from rodents are marginal or identify unusual target organ specificity. *In vivo* genotoxicity studies might assist in resolving the bioassay interpretation or aid in demonstrating activity in the target site through analysis of target organ explants.
2. Reproductive studies may have indicated the gonads as target organs for toxicity and *in vivo* germ cell mutagenicity data may be warranted to address concerns for heritable genetic risk.

*In vivo* genetic toxicology studies can be conducted as stand-alone assays or may be incorporated into planned acute or subchronic toxicology studies. There are advantages and limitations associated with both approaches. In a few instances the tests may involve the use of human cell explants from exposed individuals.

### Stand-Alone Techniques

The list of tests used as stand-alone assays is the most comprehensive and consists of the most widely used assays. The tests can be applied as direct followup from their respective *in vitro* analogs. Direct followup of endpoints is the most reliable manner in which to apply this set of tests. Table 4.7 lists a set of *in vitro* tests and their followup *in vivo* models. The *in vivo* tests are listed in order of scientific preference.

Standard protocols (EPA and/or OECD guidelines) exist for most of the *in vivo* models.

TABLE 4.7
*In Vitro* Assays with *In Vivo* Follow-up Studies Measuring Comparable Endpoints

| *In vitro* assays | Recommended *in vivo* follow-up models for *in vitro* tests[a] |
|---|---|
| Gene mutation endpoint | Gene mutation endpoint |
| Ames (or other bacterial) | Mouse somatic cell coat color assay (spot test) |
| Mouse lymphoma | *Drosophila* sex-linked recessive lethal test |
| CHO (HGPRT) | Mouse specific-locus assay or suitable dominant mutation assay (germ cell)* |
| Yeast forward or reverse mutation | |
| Chromosome aberration endpoint | Chromosome aberration endpoint |
| *In vitro* cytogenetic analysis in various cell lines | Rodent micronucleus |
| | Rodent bone marrow metaphase analysis |
| | Dominant lethal assay (germ cell)* |
| | Heritable translocation (germ cell)* |
| | Germ cell chromosome aberrations* |
| UDS in primary hepatocytes | Rat *in vivo/in vitro* UDS in hepatocytes |
| Sister chromatid exchange (SCE) | *In vivo* SCE in bone marrow or other cell types |
| *In vitro* adduct formation (DNA binding) | DNA binding in selected target organs using radiolabeled chemical or $^{32}$P-Postlabeling* |
| *In vitro* cell transformation | Liver focus assay in rats |

[a] Asterisk indicates test or technique that may be applicable to the development of heritable genetic risk estimates.

The general interpretation applied to the use of *in vivo* data is that they may take precedence over *in vitro* results if (a) the design of the *in vivo* study is adequate to detect weak responses and (b) evidence can be provided that the test article and/or its metabolites have access to the target site.

## Expansion of Standard Toxicology Assays

These *in vivo* models are particularly valuable in developing an *in vivo* data base from core toxicology studies. Most of the effort associated with dosage determination, compound administration, and animal care will be incorporated into the toxicology study design. The genetic toxicology study requires (a) the addition of a few animals or (b) removal of specific cells from existing core treatment and control animals.

Data from these studies may be used to develop early hazard assessment for major products or to study target organ and metabolism

## TABLE 4.8
## Examples of *In Vivo/In Vitro* Techniques Used to Identify Genotoxic Chemicals

| Method | Target cell employed | Genetic endpoint assessed | Reference |
|---|---|---|---|
| Expose animals *in vivo*, collect and culture *in vivo* blood; cells arrested in metaphase are evaluated for structural and numerical chromosome damage | Circulating lymphocytes[a] | Clastogenicity | 50 |
| Same as above, except that bromodeoxyuridine (BrdU) is administered to the *in vitro* cultures prior to metaphase arrest; sister chromatid exchange (SCE) is evaluated in M2 cells | Circulating lymphocytes[a] | Direct chemical/DNA interaction | 51,52 |
| Expose animals *in vivo*, collect blood, and place aliquots into selective and nonselective *in vitro* culture medium; proliferation of cells in selective medium following stimulation by growth factors is indication of mutation | Circulating T-lymphocytes[a] | Gene mutation at the HGPRT or other suitable locus | — |
| Expose animals *in vivo*, sacrifice animals, perfuse liver, and collect hepatocytes; hepatocytes are cultured *in vitro* with $^3$H-thymidine and evaluated by autoradiographic method | Primary hepatocytes[a] | Stimulation of DNA repair system following induced damage; incorporation of $^3$H is indication of repair synthesis | 53 |
| Expose animals *in vivo*, collect urine; urine can be tested directly as a mutagen in an *in vitro* assay; urine may also be concentrated and/or deconjugated prior to use *in vitro* | *In vitro* assays ranging from bacteria to cultured rodent or human cells | Only indicates that genotoxic agent has been formed *in vivo* and excreted via bladder | 54 |
| Expose animals *in vivo*, collect sperm by surgical methods or induced ejaculation; fix and stain sperm for microscopic examination; frequency of abnormal spermheads recorded | Mature sperm | Genetic basis of this effect not well understood | 55 |
| Expose pregnant females *in vivo*, sacrifice, collect, and dissociate fetal cells; culture fetal cells in soft agar and score for colony formation | Fetal cells | Anchorage-independent cell growth *in vitro* is a characteristic of transformed cells (e.g., clone primary cells in soft agar) | — |

[a] Other target cells can be used, but collection of cells other than circulating blood required rather severely invasive methods that are not usually applied on a routing basis (e.g., liver bone marrow or skin biopsies; lung lavage or peritoneal exudate).

questions that have arisen regarding the test material. The extent of these studies is similar to the previous group, since many of the same endpoints are included. The major limitations are that, occasionally, dose levels or routes of administration established for acute and/or subchronic toxicology studies might not be applicable for genetic evaluations. Occasionally, species selection may prevent successful performance of genetic studies.

Table 4.8 lists a set of *in vivo* endpoints that can be conveniently tied to acute and 14-day, 28-day, and 90-day subchronic toxicology studies in mice and rats.

While this group of tests may be viewed more as research and development tools, if properly constructed their results may be interpreted in the same fashion as the results from stand-alone assays with respect to confirming or balancing *in vitro* results.

### Regulations Affecting Testing

Genetic toxicology studies submitted to U. S. or international regulatory agencies must comply with certain regulations and testing guidelines. In the United States the FDA and EPA Good Laboratory Practices regulations determine performance of tests and the documentation required for reporting the results.[56] In addition, there are employee health and safety requirements and facility specifications incumbent upon most commercial testing operations.[57]

Several government institutions and advisory bodies have published standard study designs for most genetic toxicology assays.[23,39,40] Some regulatory agencies insist that all reports follow guideline protocols from one of these sources. Berry and Litchfield have prepared a review of the regulatory requirements for mutagenicity testing.[1] The requirements are similar in the United States, Canada, Japan, and Western Europe.

## DATA ANALYSIS AND INTERPRETATION

### Primary Evaluation of Individual Tests

Each type of test system is amenable to some form of data analysis. Most are evaluated mathematically to determine the statistical significance of the results. A review of statistical analyses for short-term tests is given by Ehrenberg.[58] Those tests not subject to critical statistical analysis are generally evaluated by preestablished operational criteria. This usually involves setting a level (e.g., multiple of the control) below which the results are considered negative and above which they are called positive. Most criteria are based on laboratory experience or communication among investigators.[59] Criteria described in EPA, OECD, and EEC guidelines are not specific as to type of analysis to be used, although statistical methods are recommended. A careful review of published data, such as those derived from the Gene-Tox program and other large collaborative studies, will be necessary for standardization of data analysis.

### Evaluation Approaches for Batteries

The second level of data evaluation leads to a description or conclusion regarding the biological activity of a test substance. Is the compound a mutagen or a clastogen? Does exposure to the material constitute a health hazard? This level of data analysis is more elusive than the primary evaluation because it is based on a limited experience and requires intuitive processes. Therefore, most interpretive evaluations are qualitative and only define hazard "potential."

The three most common methods for evaluating results from screening batteries are:

1. *The decision tree method*, in which results from one test direct the investigator to the next test or to an interpretation or categorization of the chemical.[60]
2. *The critical mass approach*, in which a specified number of positive tests in a given category automatically defines the category of a test material, regardless of the spectrum of response or the concentration needed to achieve the response. Negative responses are not contributory unless the results are negative in all tests conducted.
3. *The weight of evidence approach*, in which each test has a specific weight proportional to its reliability in defining a true positive or negative effect. The overall evaluation is based on the trend of responses and is useful for mixed results, since both positive and negative data can be accommodated.

### The Ideal Test Battery

The first action required to initiate a screening program in genetic toxicology is selection of a test battery. Some parameters of test selection have been described previously. The following information is more specific about the requirements for testing. There is probably not sufficient information in the form of test data on mutagens and nonmutagens available in the literature to construct a universal battery for the detection of genotoxicity. The only generalizations that seem to be supportable are as follows:

1. The tests included in a battery should cross several phylogenetic levels, including at least prokaryotic and eukaryotic cell types.
2. The combination of tests should detect more than one type of genetic endpoint and should include one test measuring a generalized mechanism of genotoxicity, such as DNA repair, SCE, or cell transformation.
3. *In vivo* as well as *in vitro* tests should be included in a comprehensive test battery.
4. *In vitro* tests should be amenable to the detection of active metabolites produced by enzymatic bioactivation of test agents.

More advantageous use of individual tests requires a knowledge of how each test predicts either germ cell mutation or tumor development in animal models. A test battery that has been optimized for rodent carcinogen screening may not be suited for germ cell mutagen screening. Several methods have been used to establish the optimal tests for predicting rodent carcinogens. At the present time none appear reliable enough to be recommended with confidence. The Ames test plus two or three mammalian cell *in vitro* assays seems to be a generally accepted battery for carcinogen identification. A recent report from the EPA Gene-Tox program compared a wide range of short-term assays for prediction of germ cell damage.[61] The report indicated that the mouse spot test, unscheduled DNA synthesis *in vivo*, and the micronucleus test were good predictors for the mouse specific–locus assay and that unscheduled DNA

TABLE 4.9
The Application of General Genetic Testing Methods to Carcinogen and Mutagen Assessment. Genetic Testing Categories I–III[a]

| Primary type of identification obtained | I. *In vitro* detection of mutation induction and chromosome damage | II. *In vivo* detection of mutation induction and chromosome damage | | III. *In vitro*/*in vivo* detection of genotoxic damage not leading to somatic or gonadal heritable alterations |
|---|---|---|---|---|
| | | Somatic | Gonadal | |
| Identification of presumptive animal carcinogens | A | A | NA | A |
| Identification of presumptive animal mutagens | A | A | A | (NA) |
| Risk analysis of heritable genetic effects | NA | NA | A | NA |
| Risk analysis for carcinogenic effects | NA | (A) | NA | NA |
| Identification of animal mutagens | NA | A | A | NA |

[a] From Brusick.[62] A, Applicable; NA, not applicable; parentheses signify special circumstances or explanation.

TABLE 4.10
Summary of Possible Roles for Specific Short-Term Tests in Chemical Hazard Assessment[a]

| General assay type | Testing category from Table 4.9 | Role: ICP | IMP | EA | RA |
|---|---|---|---|---|---|
| Microbial assays | | | | | |
| Ames Test | I | + + | + | + | NA |
| $WP_2$ and derivatives | I | + | + | NA | NA |
| Bacterial repair tests | III | + | NA | NA | NA |
| Yeast mutation | I | + | + + | NA | NA |
| Yeast mitotic recombination | III | + | + | NA | NA |
| *In vitro* mammalian cell assay | | | | | |
| Mouse lymphoma assay | I | + | + + | NA | NA |
| CHO or V79 mutation assay | I | + | + + | NA | NA |
| UDS assay | III | + + | NA | NA | NA |
| Chromosome aberrations | I | + | + + | + | NA |
| Sister chromatid exchange | I/III | + + | NA | + | NA |
| Cell transformation | III | + + | NA | NA | NA |
| *In vivo* mammalian assays | | | | | |
| Dominant lethal assay | II | NA | + + | NA | + |
| Cytogenetic analysis | II | + | + | + | + + |
| Micronucleus assay | II | + | + | NA | (+) |
| Spermhead abnormality assay | II/III | NA | (+) | NA | (+) |
| Heritable translocation assay in mice | II | NA | + | NA | + + |
| Specific-locus assay in mice | II | NA | + + | NA | + + |
| DNA adduct formation | III | (+) | (+) | NA | + + |
| *In vivo* submammalian assay | | | | | |
| *Drosophila* assays | II | + | + + | (+) | (+) |
| Plant cytogenetics | II | NA | (+) | + | NA |

[a] From Brusick.[62] ICP, Identifies mammalian carcinogenic potential; IMP, identifies mammalian mutagenic potential; EA, environmental assessment application; RA, risk assessment application. +, Applicable; + +, >+; NA, not applicable; (+), possible application under some conditions.

synthesis *in vivo*, dominant lethal test, SCE *in vitro*, sperm abnormalities, and the *Drosophila* heritable translocation assay appeared good predictors for the mouse heritable translocation assay.

Tables 4.9 and 4.10 summarize, in general, the appropriate application of various tests to either carcinogen or mutagen batteries.

## Interpretation of Results from Test Batteries

The ultimate application of test results is the assessment of possible hazard for humans exposed to the test agent. In this context, hazard assessment is defined as the most likely toxic outcome for humans exposed acutely or chronically to biologically significant levels of the agent. Hazard assessments are often expressed in degrees of concern rather than in expected increases in disease incidence as might be done for risk analysis.

Hazard assessments based on single tests, especially *in vitro* assays, are less reliable than those based on a battery of tests, especially if the previous recommendations for construction of the battery are followed.

Because the response profile of an agent subjected to a test battery is seldom uniformly negative or uniformly positive across all the tests, the evaluation approaches described earlier will be required to handle the mixed responses.

Two approaches will be described as examples of possible courses one might follow in interpreting data from test batteries (i.e., multitest response matrices). Each of the three types of test batteries described earlier in this chapter is amenable to interpretation in a systematic manner using one of the two approaches.

The first method is a two-dimensional analysis of data, which establishes estimates of hazard on the basis of positive test results. Estimates are derived from the assumption that tests measuring gene mutation and/or chromosome damage are directly responsible for, and therefore more relevant to human disease than, other genotoxic endpoints. Chemicals that induce only primary DNA damage, for example, are not accorded the same degree of concern because the etiology of human diseases cannot be linked to nonmutational endpoints. Consequently, under the proposed approach to hazard assessment, test systems stratify into three classes indicating genotoxic hazard according to their position in the genotoxic pathway outlined in Chapter 2 (Figure 2.22).

Class 3. Assays that measure gene mutation or structural and numerical chromosome aberrations.
Class 2. Assays that measure SCE, recombination, DNA-repair synthesis, DNA-strand breakage, or DNA cross-linking.
Class 1. Assays that measure adduct formation or inhibition of DNA synthesis or replication.

Class 1 represents the lowest degree of concern and Class 3 the highest.

The second dimension of this analysis stratifies test organisms by their phylogenetic relatedness to humans. Data from a test battery that includes at least one test from each of the three endpoint classes are analyzed using the scale in Table 4.11.

The information in Table 4.11 can be made semiquantitative by placing it in a two-way evaluation table (Table 4.12). In this table, there are

## TABLE 4.11
## Two-Dimensional Analysis of the Relevance of Test Results

| | | |
|---|---|---|
| Most | Tests that detect DNA alterations in somatic cells of humans exposed to genotoxic agents | Evidence from human studies for the induction of somatic and germ cell, gene, or chromosome mutations |
| | Tests that detect alterations in DNA synthesis and replication or tests that measure direct DNA damage and repair *in vivo* using mammalian models | Tests that detect gene or chromosome mutation *in vivo* using mammalian models |
| | Tests that detect alterations in DNA synthesis and replication or tests that measure direct DNA damage and repair *in vitro* using metabolically proficient mammalian cell systems | Tests that detect gene mutation or cell transformation *in vitro* using metabolically proficient mammalian cell systems |
| | Tests that detect alterations in DNA synthesis and replication or tests that measure direct DNA damage and repair using metabolically sufficient or exogenously supplemented nonmammalian cell systems | Tests that detect mutation or chromosome damage using metabolically sufficient or exogenously supplemented nonmammalian cell systems |
| Least | | |
| | Relevance of the genotoxic endpoint detected | |

## TABLE 4.12
## Data Analysis Table for Categorizing Hazard from Positive Test Results

| Stratification of assays by complexity of bioassay system: Assay relevance | | | Least relevant indicator of somatic mutation | | Most relevant indicator of somatic mutation |
|---|---|---|---|---|---|
| | | | Class 1 | Class 2 | Class 3 |
| Least | Plant and prokaryotic systems | A | 1A | 2A | 3A |
| | Lower eukaryotic microorganisms and insect systems | B | 1B | 2B | 3B |
| | Mammalian *in vitro* systems | C | 1C | 2C | 3C |
| | Mammalian *in vivo* systems | D | 1D | 2D | 3D |
| Most | Human *in vivo* data | E | 1E | 2E | 3E |

TABLE 4.13
A Proposed Scheme for Classifying Chemicals on the Basis of Test Results

| Chemical classification | Response required for classification of hazard potential | Implications for oncogenesis |
|---|---|---|
| Genotoxic agent | Positive response in at least one square | None, unless other conditions defined below are met |
| Mutagenic agent | Positive response in at least one class 3 square | Carcinogenic potential is possible and the level of probability increases from 3A to 3D tests |
| Animal cell somatic mutagen[a] | Positive responses in at least one test each from 1C and 2C and at least two tests from 3C | Relevance of test systems, correlation coefficients, and nature of the DNA lesions all point toward carcinogenic activity |
| Probable animal (human) mutagen[b] | Positive responses in at least two tests each of 2C and 3C plus a positive response in 2D, 3D, or 2E | Relevance of test systems, correlation coefficients, nature of the DNA lesions, and *in vivo* chemical pharmacodynamics all point toward *in vivo* tumorigenicity |
| Human mutagen | Positive results or evidence in 3E | Probably increase of risk of tumor induction if exposure persists |

[a] The test material has biological properties in common with materials known to produce cancer in experimental animals, but the pharmacodynamics of *in vivo* exposure is not known.
[b] The test material has relevant genotoxic activity in mammalian cells and also has been shown to express that activity under *in vivo* test conditions.

five phylogenetic strata, ranging from prokaryotic and plant cells to human *in vivo* responses, and the three degrees of data information (Classes 1, 2, 3).

A tested material may provide data from as few as three tests or as many as ten or more. A proposed agent classification scheme is outlined in Table 4.13. An interpretation of the multitest is derived from combinations of numbers and letters associated with positive responses. The table uses combinations of the most relevant data acquired to classify the chemicals according to their potential hazard. In many cases, the data available on a chemical will not be sufficient to extend its classification of a chemical beyond "genotoxic" or "mutagenic"; in a few cases, a chemical may be defined as a human hazard. This approach is similar in many respects to the EPA mutagenic assessment guidelines.[8]

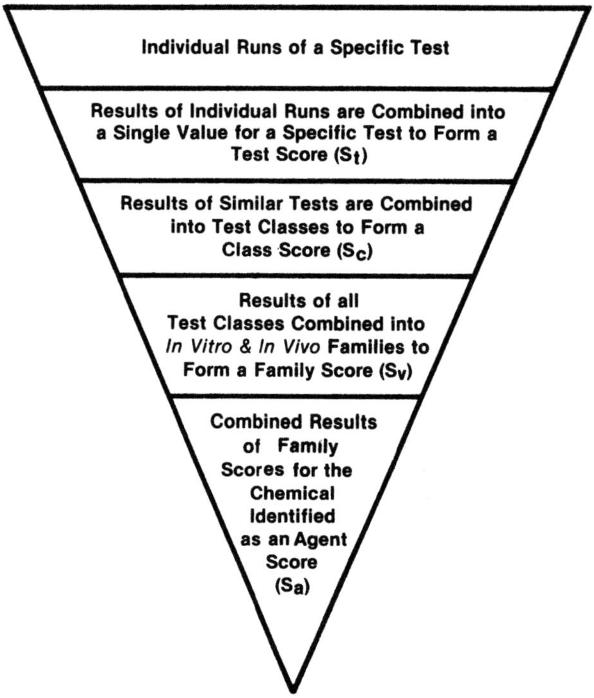


FIGURE 4.8. General process of data reduction. During this process individual entrants are modified by a series of factors and combined into a weighted mean. Each step combines responses from similar tests and eventually merges *in vitro* and *in vivo* responses. The final score defined as an $S_a$ (agent score) represents the weighted average score for all tests included in the assessment.

The second approach is a "weight-of-evidence" method recently proposed by ICPEMC Committee 1.[5] It differs substantially from the first method in that it relies upon both positive and negative responses from tests. Although the method is quantitative, it uses intuitive processes in the analysis. A weight (value) is set for each test, and other factors (e.g., active dose, replication, need for activation) that modulate the test weight are serially multiplied by the value of the test. A ratio of the modified weight to the unmodified weight forms the basis for the test score ($S_t$). This process is repeated for each test in the battery, and the results are then combined and weighted over a sequence of data reduction steps, eventually leading to a single agent score ($S_a$) (Figure 4.8).

Using this analysis, each of the test results would contribute positively or negatively to the net agent score. As the data are merged into composite scores the contribution of each entry is retained and illustrated

TABLE 4.14
Identification of the Test Types Included in the *in Vitro* and *in Vivo* Families

| *In vitro* family ($S_{f1}$) | *In vivo* family ($S_{f2}$) |
|---|---|
| Class 1: Prokaryotic—primary DNA damage | Class 1: Mammalian—UDS |
| Class 2: Lower eukaryotic—primary DNA damage | Class 2: Insect—*Drosophila*-gene mutation |
| Class 3: Mammalian cell—UDS | Class 3: Mammal—somatic spot test |
| Class 4: Prokaryotic—gene mutation | Class 4: Mammal—specific-locus assay |
| Class 5: Lower eukaryotic—gene mutation | Class 5: Mammal somatic SCE |
| Class 6: Mammalian cell—gene mutation | Class 6: Mammal—somatic micronuclei |
| Class 7: Lower eukaryotic—aneuploidy | Class 7: Mammal—somatic chromosome aberration |
| Class 8: Mammalian cell—SCE | Class 8: Mammal—dominant lethal |
| Class 9: Mammalian cell—aberrations | Class 9: Mammal—heritable translocation |
| Class 10: Mammalian cell—transformation | Class 10: Mammal—germ cell chromosome aberration |
| | Class 11: Mammal—sperm morphology |

to permit the user to see the response profile throughout the data reduction process (Figure 4.9 and Table 4.14).

The advantages of this approach are that it tends to simulate the type of intuitive logic used by expert genetic toxicologists in evaluating complex data sets. The approach is also amenable to validation and calibration against an external standard such as *in vivo* carcinogen ranking.

Both of the data analysis stems are potentially valuable in assessing and interpreting multitest data bases. The weight-of-evidence approach may prove to be the more beneficial in that it is more quantitative, it can be modified on the basis of validation data sets, and it defines activity on the basis of all data rather than on selected tests or only positive data.

## SUMMARY

In summary, short-term tests can be useful for more than just measuring the mutagenicity of a compound *per se*. They are important in decisions on utilization of resources for other types of toxicologic tests that require a major investment. Short-term tests are also capable of determining the potential hazard of compounds that are going to the marketplace, but are not required to be subjected to a full-scale toxicologic assessment. Other uses involve occupational health considerations. Chemical process intermediates may never reach the marketplace, but

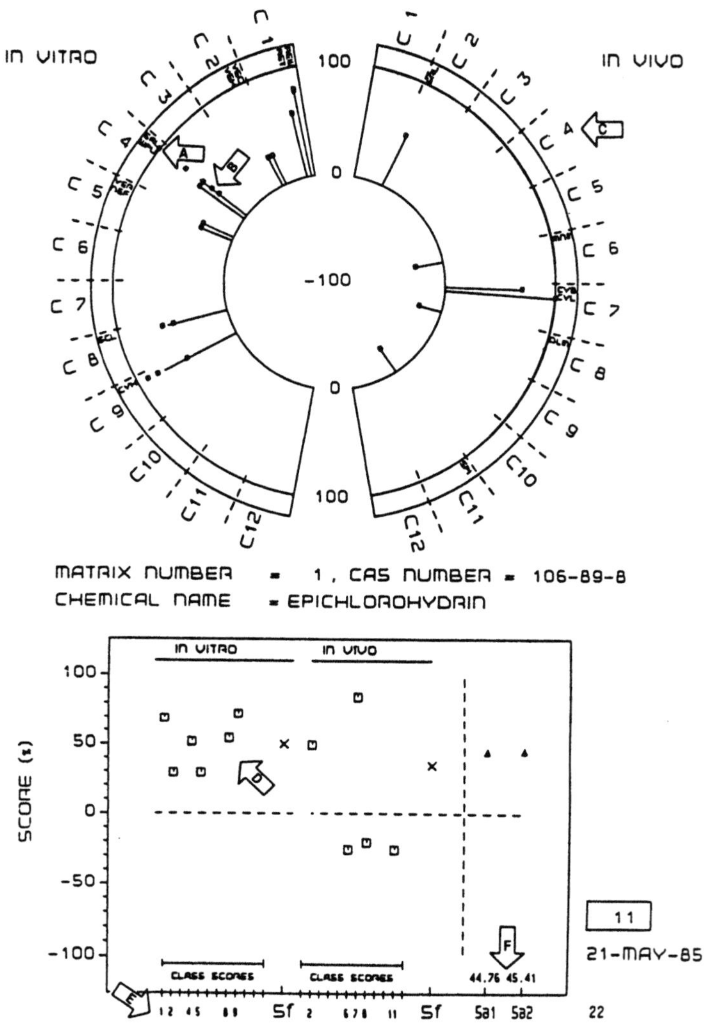


FIGURE 4.9. Output from ICPEMC-type battery analysis. The upper portion of the figure, showing the two hemispheres, shows ten classes for *in vitro* data (left hemisphere) and 11 classes for *in vivo* data (right hemisphere). If a test has been conducted from a given class, then (A) a three-letter code identifying the specific test (e.g., Sal, Ames *Salmonella* mutation assay; WPU, reverse mutation in *E. coli*) and (B) a point (small squares), representing the test score $S_t$, identify each trial conducted. The points, if positive, range from zero to $+100$; if negative, from 0 to $-100$. The direction and length of the line connecting the points represent the weighted mean of all trials. (C) The classes represented are identified by the numbered code in Table 4.14. The lower portion of the figure plots the class scores. (D) Each class score $S_c$ (□) represents the weighted means of each test type included in that class (e.g., under class C4 of the *in vitro* hemisphere the data show results from four *Salmonella* assays and one *E. coli* $WP_2$ *uvr*A− assay). (E) The identity of each class is shown

there will be certain individuals who are exposed to them and it will be important to know, from an occupational hazard standpoint, whether there is any risk to employees exposed to these intermediates.

Short-term tests identify potential for genotoxicity. Once potential has been established, the economics of development and use pattern of the compound will determine the subsequent steps in toxicologic safety evaluation prior to marketing.

In the near future, short-term test results may find greater application in hazard analysis, especially in situations that preclude the use of large-scale animal bioassays, via scoring systems that integrate complex data from several tests into a value that represents the genotoxic hazard of the tested material. One can hope that these methods will reduce our dependence on the "looking for positives" philosophy that has dominated genetic screening.

## REFERENCES


1. Berry, D. J., and Litchfield, M. H.: A review of the current regulatory requirements for mutagenicity testing. In *Progress in Mutation Research*, Vol. 5 (J. Ashby and F. J. de Serres, eds.), Elsevier, Amsterdam, pp. 727–740, 1985.
2. Butterworth, B. E., and Golberg, L. (eds): *Strategies for Short-Term Testing for Mutagens/Carcinogens*, CRC Press, West Palm Beach, Florida, 1979.
3. Bridges, B. A.: Some general principles of mutagenicity screening and a possible framework for testing procedures. *Environ. Health Perspect.* **6**:221, 1973.
4. Flamm, W. G.: A tier system approach to mutagen testing. *Mutat. Res.* **26**:329, 1974.
5. ICPEMC Committee 1 Final Report. Screening strategy for chemicals that are potential germ-cell mutagens in mammals. *Mutat. Res.* **114**:117–177, 1983.
6. Komatsu, H., Kabizoe, T., Niijima, T., Kawachi, T., and Sugimura, T.: Increased sperm abnormalities due to dietary restriction. *Mutat. Res.* **93**:439–446, 1982.
7. Soares, E. R.: Genetic aspects of short-term testing and an appraisal of some *in vivo* mammalian test systems. In *Strategies for Short-Term Testing for Mutagens/Carcinogens* (B. E. Butterworth and L. Goldberg, eds.), CRC Press, West Palm Beach, Florida, pp. 67–76, 1979.
8. Proposed guidelines for mutagenicity risk assessment. *Fed. Reg.* **49**(Part IX):46314–46321, 1984.
9. Ashby, J.: Screening chemicals for mutagenicity: Practices and pitfalls. In *Mutagenicity: New Horizons in Genetic Toxicology*, Academic Press, New York, pp. 1–33, 1982.


---

by the number along the base of the graph; as given in Table 4.14. The weighted means for each family of classes ($S_v$) are represented by ×, one × for *in vitro* and one for *in vivo*. (F) Finally, the $S_{a1}$ score (▲) represents the weighted mean of the two family scores and the $S_{a2}$ (▲) score the unweighted mean of all class scores from both families. The values of the $S_a$ scores represent an estimate of the overall genotoxicity for that chemical, including input from each trial of every test performed. Both parts of the figure show how mixed data are handled.

10. Brusick, D., and Auletta, A.: Developmental status of bioassays in genetic toxicology. *Mutat. Res.* **153**:1–10, 1985.
11. Rundell, J. O.: *In vitro* cell transformation, An overvciew. In *Carcinogenesis and Mutagenesis Testing* (J. F. Douglas, ed.), Humana Press, Clifton, New Jersey, pp. 39–62, 1984.
12. Mishra, N. K., and DiMayorca, G.: *In vitro* malignant transformation of cells by chemical carcinogens. *Biochim. Biophys. Acta* **355**:205–219, 1974.
13. Pienta, R. J., Poiley, J. R., and Lebherz III, W. B.: Morphological transformation of early passage golden Syrian hamster embryo cells derived from cryopreserved primary cultures as a reliable *in vitro* bioassay for identifying diverse carcinogens. *Int. J. Cancer* **19**:642–655, 1977.
14. Callen, D. F., and Philpot, R. M.: Cytochrome P-450 and the activation of promutagens in *Saccharomyces cerevisiae. Mutat. Res.* **45**:309–324, 1977.
15. Vogel, E., and Sobels, F. H.: The function of *Drosophila* in genetic toxicology testing. In *Chemical Mutagens: Principles and Methods for Their Detection*, Vol. 4 (A. Hollaender, ed.), Plenum Press, New York, pp. 93–142, 1976.
16. Gabridge, M. G., and Legator, M. S.: A host-mediated microbial assay for the detection of mutagenic compounds. *Proc. Soc. Exp. Biol. Med.* **130**:831–834, 1969.
17. Arni, P., Mantel, T., Deparade, E., and Muller, D.: Intrasanguine host-mediated assay with *Salmonella typhimurium. Mutat. Res.* **43**(3):291–307, 1977.
18. Speck, W. T., Stein, A. B., and Rosenkranz, H. S.: Mutagenicity of metronidazole: Presence of several active metabolites in human urine. *J. Natl. Cancer Inst.* **56**:283–284, 1976.
19. Legator, M. S., Connor, T., and Stoeckel, M.: Detection of mutagenic activity of metronidazole and miridazole in body fluids of humans and mice. *Science* **188**:1118–1119, 1975.
20. Gletten, F., Weekes, U., and Brusick, D.: *In vitro* metabolic activation of chemical mutagens. I. Development of an *in vitro* mutagenicity assay using liver microsomal enzymes for the activation of dimethylnitrosamine to a mutagen. *Mutat. Res.* **28**:113–122, 1975.
21. Malling, H. V.: Dimethylnitrosamine formation of mutagenic compounds by interaction with mouse liver microsomes. *Mutat. Res.* **13**:425, 1971.
22. Ames, B. N., Durston, W. E., Yamasaki, E., and Lee, F. D.: Carcinogens are mutagens. *Proc. Natl. Acad. Sci. USA* **72**:979, 1973.
23. OECD (Organization for Economic Co-operation and Development): *Guidelines for Testing of Chemicals*, 1981; revised May 1983.
24. Stout, D. L., and Becker, F. F.: Metabolism of 2-aminofluorene and 2-acetylaminofluorene to mutagens by rat hepatocyte nuclei. *Cancer Res.* **39**(4):1168–1173, 1979.
25. Huberman, E., and Sachs, L.: Cell-mediated mutagenesis with chemical carcinogens. *Int. J. Cancer* **13**:326, 1974.
26. De Serres, F., Fouts, J. R., Bend, J. R., and Philpot, R. M. (eds.): *In Vitro Metabolic Activation in Mutagenesis Testing*, Elsevier/North-Holland, Amsterdam, 1976.
27. Malling, H. V., and Frantz, C. N.: *In vitro* versus *in vivo* metabolic activation of mutagens. *Environ. Health Perspect.* **6**:71–82, 1973.
28. Tang, T., and Friedman, M. A.: Carcinogen activation by human liver enzymes in the Ames mutagenicity test. *Mutat. Res.* **46**:387–394, 1977.
29. Haworth, S., Lawlor, T., Mortelmans, K., Speck, W., and Zeiger, E.: *Salmonella* mutagenicity test results for 250 chemicals. *Environ. Mutagen.* (Suppl.)**1**:3–142, 1983.
30. Brusick, D. J.: *In vitro* mutagenesis assays as predictors of chemical carcinogens in mammals. In *Toxicology Annual*, Vol. 2 (C. L. Winek, ed.), Marcel Dekker, New York, pp. 79–109, 1977.

31. Brusick, D. J.: Mutagenicity and carcinogenicity correlations between bacteria and rodents. In *Cellular Systems for Toxicity Testing 407* (G. M. Williams, V. C. Dunkel, and V. A. Ray, eds.), New York Academy of Sciences, New York, pp. 164–176, 1983.
32. Squire, R. A.: Ranking animal carcinogens: A proposed regulatory approach. *Science* **214**:877–880, 1981.
33. Ray, V. A.: Application of microbial and mammalian cells to the assessment of mutagenicity. *Pharmacol. Rev.* **30**(4):537–546, 1979.
34. Waters, M., and Auletta, A.: The GeneTox Program. *J. Chem. Inf. Comput. Sci.* **21**:35–38, 1981.
35. Pet-Edwards, J., Chankong, V., Rosenkranz, H. S., and Haimes, Y. Y.: Applications of the carcinogenicity prediction and battery selection (CPBS) method to the Gene-Tox data base. *Mutat. Res.* **153**:187–200, 1985.
36. Weisburger, J. H., and Williams, G. M.: The decision point approach for systematic carcinogen testing. *Food Cosmet. Toxicol.* **19**:561–566, 1981.
37. Ashby, J., de Serres, F. J., Draper, M., Ishidate, M., Margolin, B., Matter, B., and Shelby, M. D.: *Short-Term Tests for Carcinogens, Results of the IPCS Study*, Elsevier, Amsterdam, 1985.
38. Heinze, J. E., and Poulsen, N. K.: The biological independence of several genetic toxicology tests: Implications for the design of testing batteries. Presented at the 12th Annual Meeting of the Environmental Mutagen Society, Boston, Massachusetts, February 27, 1982.
39. EPA (Environmental Protection Agency): Office of Pesticides and Toxic Substances, Health Effects Test Guidelines, 1982, EPA Publication 560/682-001, National Technical Information Service, Springfield, Virginia. Revised 1985.
40. EEC (European Economic Community): Official Journal of the European Communities, 6th Amendment to Directive 67/548/EEC, Annex VII, 15.10.79, and Annex V, EEC Directive 79-831, Part B, Toxicological Methods of Annex VIII, Draft, July 1983.
41. Brusick, D. J.: Evaluation of "Genotoxic effects in cultured mammalian cells produced by low pH treatment conditions and increased ion concentrations." *Environ. Mutagen.* **8**:879–886, 1986.
42. Cifone, M. A.: Relationship between increases in mutant frequency in L5178Y TK +/− mouse lymphoma cells at low pH and metabolic activation. *Environ. Mutagens* **1**(Suppl. 3):27, 1985.
43. Thilagar, A.: Effect of low pH caused by glacial acetic acid and hydrochloric acid on the chromosome aberrations in CHO cells. *Environ. Mutagen.* **8**:874–886, 1986.
44. Nanni, N., Bauer, C., Cundari, E., Corsi, C., Del Carratore, R., Nieri, R., Paolini, M., Crewshaw, J., and Bronzetti, G.: Studies of genetic effects in the D7 strain of *Saccharomyces cerevisiae* under different conditions of pH. *Mutat. Res.* **139**:189–192, 1984.
45. Cole, J., Muriel, W. J., and Bridges, B. A.: The mutagenicity of sodium fluoride to L5178Y [wild-type and TK+/− (3.7.2c) mouse lymphoma cells]. *Mutagenesis* **1**:157–167, 1986.
46. Ishidate, M., Jr., Sofuni, T., Yoshikawa, K., Hayashi, M., Nohmi, T., Sawada, M., and Matsuoka, A.: Primary mutagenicity screening of food additives currently used in Japan. *Food Chem. Toxicol.* **22**:623–636, 1984.
47. Generoso, W. M., Cain, K. T., Huff, S. W., and Gosslee, D. G.: Heritable translocation test in mice. In *Chemical Mutagens: Principles and Methods for Their Detection*, Vol. 5 (A. Hollaender and F. J. de Serres, eds.), Plenum Press, New York, pp. 55–77, 1978.
48. Würgler, F. E., Graf, U., and Berchtold, W.: Statistical problems connected with the sex-linked recessive lethal test in *Drosophila melanogaster*. I. The use of the Kastenbaum–Bowman test. *Arch. Genet.* **48**:158, 1975.
49. ICPEMC, Committee 4 Final Report. Estimation of genetic risks and increased incidence of genetic disease due to environmental mutagens. *Mutat. Res.* **115**:225–291, 1983.

50. Preston, R. J., and Brewen, J. G.: X-ray-induced chromosome aberrations in the leucocytes of mouse and man. In *Mutagen-Induced Chromosome Damage in Man* (H. J. Evans and D. C. Lloyd, eds.), Yale University Press, New Haven, Connecticut, pp. 30–40, 1978.
51. Allen, J. W., Shuler, C. F., Mendes, R. W., and Latt, S. A.: A simplified technique for *in vivo*, analysis of sister-chromatid exchanges using 5-bromodeoxyuridine tablets. *Cytogenet. Cell Genet.* **18**(4):231–237, 1977.
52. Kligerman, A. D., Wilmer, J. L., and Erexson, G. L.: Characterization of a rat lymphocyte culture system for assessing sister chromatid exchange after *in vivo* exposure to genotoxic agents. *Environ. Mutagen.* **3**(5):531–543, 1981.
53. Mirsalis, J. C., and Butterworth, B. E.: Detection of unscheduled DNA synthesis in hepatocytes isolated from rats treated with genotoxic agents: An *in vivo–in vitro* assay for potential carcinogens and mutagens. *Carcinogenesis* **1**:621–625, 1980.
54. Durston, W. E., and Ames, B. N.: A simple method for the detection of mutagens iin urine: Studies with the carcinogen 2-acetylaminofluorene. *Proc. Natl. Acad. Sci. USA* **71**(3):737–741, 1974.
55. Wyrobek, A., and Bruce, W. R.: Chemical induction of sperm abnormalities in mice. *Proc. Natl. Acad. Sci. USA* **72**:4425–4429, 1975.
56. Department of Health, Education and Welfare, Food and Drug Administration: Good laboratory practice regulations for nonclinical laboratory studies. *Fed. Reg.* **43**(Part II):59986–60020, 1978.
57. Nemchin, R. G., and Brusick, D. J.: Basic principles of laboratory safety. *Environ. Mutagen.* **7**(6):947–970, 1985.
58. Ehrenberg, L.: Aspects of statistical inference in testing for genetic toxicity. In *Handbook of Mutagenicity Test Procedures* (B. J. Kilbey, M. Legator, W. Nichols, and C. Ramel, eds.), Elsevier, Amsterdam, pp. 419–459
59. de Serres, F. J., and Shelby, M. D.: The *Salmonella* mutagenicity assay: Recommendations. *Science* **203**:563–565, 1979
60. Food Safety Council: *Proposed System for Food Safety Assessment*, 1980.
61. Russell, L. B., Aaron, C. S., deSerres, F., Generoso, W. M., Kannon, K. L., Shelby, M., Springer, L., and Voytele, P.: Evaluation of mutagenicity assays for purposes of genetic risk assessment. *Mutat. Res.* **134**:143–157, 1984.
62. Brusick, D. J.: New developments in short-term tests for carcinogenicity: Genetic toxicology testing. In *Chemical Induction of Cancer—Structural Bases and Biological Mechanisms*, Vol. IIIC (J. C. Arcos, ed.), Academic Press, New York, pp. 481–521, 1985.

# CHAPTER 5

# Genetic Risk Estimation

## INTRODUCTION

The state of the art in risk estimation for genotoxic effects in humans was probably best summarized by Newcombe[1]:

> little is known about the extent of the effect on health which a given increase in mutation rate would cause in man. This is true because of a lack of certainty about the amount of ill health that is maintained in the population by the pressure of recurrent natural mutations. Without such knowledge, the importance of exposures to known mutagens is difficult to assess quantitatively with a view to setting reasonable standards for the protection of the human gene pool.

In other words, until the precise human genetic load is determined, it will be difficult to place into perspective any estimated increase in mutations due to environmental mutagens. Many factors that would be critical to a risk assessment, such as the relative contribution of different types of genetic lesions to the risk estimate (e.g., dominant mutations, recessive mutations, chromosomal aberrations), are not well appreciated. It has been estimated that 1–2% of all liveborn humans are affected by genetic alterations resulting in hereditary disease.[2] Additional unknown contributions in the way of polygenic disorders also affect the human population (see Chapter 10). Distinguishing how many of these genetic lesions are the result of new spontaneous or environmentally induced mutations is extremely difficult.

Although direct evidence for environmental induction of mutation in humans has yet to be obtained, the ability of specific chemicals to produce mutations in rodent species can be clearly demonstrated.[3] As in other areas of toxicology, rodent species are considered to be adequate models for humans in genetic evaluation. Genetic toxicology also relies extensively on several associated disciplines of toxicology, such as pharmacology, physiology, biochemistry, and biometry, for the development of information needed for risk estimates. This will be evident in the subsequent discussion of approaches to risk assessment.

## DEFINITION OF RISK ESTIMATION

An operational definition of genetic risk estimation is "the quantitative estimate of the probable impact (on the gene pool of subsequent generations) from a specific mutagen exposure." This estimate is often expressed as individual or population risk rates.

There are several ways in which risk can be expressed.[4] A qualitative hazard assessment can be developed through evaluation of experimental data to determine whether exposure to a substance can result in adverse health effects. Hazard assessments, as described in Chapter 4, do not quantify the possible toxicity, but in situations where data needed to produce a quantitative risk estimate are absent, hazard assessments are valuable tools to express levels of concern for human health effects.

Approaches to quantitative risk analysis include: (1) estimating the dose required to double the normal background rate for the genetic endpoint measured (the doubling dose approach); (2) a calculation of new mutations of a specific class expected from a given exposure; (3) a function of or comparison to an equivalent dose of ionizing radiation, such as defined by the REC (radiation-equivalent chemical) concept; or (4) the parallelogram approach, which includes a combination of direct mutation analysis and DNA dosimetry.

Genetic risk estimates are not generally derived from human epidemiology studies, because so few reliable studies are devoted to measuring induced heritable mutation (dominant or recessive). Genetic risk associated with certain occupational and therapeutic exposures has been studied with inconclusive findings.[5,6] The most extensive analysis reported has been that associated with the progeny of victims exposed to radiation following the A-bomb blasts at Hiroshima and Nagasaki. In that evaluation, there has been very little evidence suggesting induced mutation at evaluatable loci,[7] although other expressions of damage, such as cancer and developmental toxicity, were identified. This lack of evidence in humans is not altogether unexpected, since cause–effect relationships spanning two generations are extremely difficult to establish using assessment techniques that monitor only a limited number of genes. Therefore, greater reliance has been placed on experimental animal studies of induced specific-locus mutations and chromosome alterations conducted at maximum tolerated doses. As a result of advances in biotech-

TABLE 5.1
Current Estimate of the Human Genetic Burden[a]

| Genetic basis | Affected individuals per thousand |
|---|---|
| Chromosome abnormality | 6.86 |
| Autosomal dominant gene mutation | 1.85–2.64 |
| Autosomal recessive gene mutation | 2.23–2.54 |
| Sex-linked gene mutation | 0.78–1.99 |
| Total | 11.72–14.03 |

[a] From Ref. 8.

nology, noninvasive methods for accurate measurement of DNA damage are becoming available and more emphasis will be placed on direct human screening.

## Estimates of Current Genetic Burdens in Humans

In spite of the apparent refractile potential of the human genome against mutagenic change and the inability to show induced mutagenic damage in humans, a substantial genetic burden exists in the human population (genetic burden—"genetic load"—is an estimate of the frequency of detrimental genetic disease in the human population). Estimates of the human genetic load range from 1 to 2% among new births and include chromosome abnormalities and autosomal dominant, recessive, and X-linked gene mutations (Table 5.1).

The latest compilation of identified genes in humans consists of 1611 actually confirmed and 1692 partially confirmed, for a total of 3303. Mutant forms of these alleles produce such human disorders as anemia, cataracts, deafness, cancer, hydrocephaly, polydactyly, albinism, glaucoma, muscular dystrophy, hemophilia, and forms of epilepsy. Matsunaga has published a comprehensive review of the types and prevalence of genetic disease in humans.[9]

In addition to the chromosomal anomalies and single-gene traits identified in humans, a large number of human disorders appear to be influenced by more than one gene set (polygenic traits). Estimates of the genetic disease burden caused by these factors (Table 5.2) are difficult to make because of complex inheritance patterns, which make it virtually impossible to trace the effect to genetic or nongenetic causes.

TABLE 5.2
Types of Human Disease Believed to Be Under Polygenic Control

| Syndrome | Estimated frequency (per 1000 live births)[a] |
|---|---|
| Anencephalus | 0.6 |
| Spina bifida | 0.3 |
| Hydrocephalus | 0.5 |
| Cardiac malformations | 7.0 |
| Cleft lip and/or palate | 3.0 |
| Talipes | 1.4 |
| Other malformations | 24.5 |

[a] Estimates from Japanese survey.[9]

## SECONDARY FACTORS INFLUENCING RISK ANALYSES

### Genotype

Several factors influence an organism's response to a genotoxic agent. Probably the most important is the individual's genotype. Variations in metabolic capacity are known to be under genetic influence[10] and metabolic processes are important in the formation and/or detoxification of mutagenic agents. Other genetically controlled responses include DNA repair and immunoresponsiveness. Specific genotypes may therefore increase the risk to mutagens by enhancing the total systemic dose caused by metabolic processes favoring the formation or transport of the biologically active agents.

### DNA Repair Capacity

The effects of genotype variants affecting DNA repair processes on the risk to somatic cell disease are well known.[11] While xeroderma pigmentosum (XP) and Bloom's and other syndromes represent extremes in mutagen sensitivity, other, more subtle effects are probably uniformly distributed throughout the human population. Generoso *et al.*[12] showed that postzygotic repair differences occur in mouse strains, and similar variations in susceptibility might also be expected for germ cells of persons with reduced or impaired DNA repair. Consequently, risk estimates derived using data from genetically homogeneous animal models may not accurately reflect risk to genetically heterogeneous human populations.

## Sex

Genetic risk is also governed by the distribution of sex chromosomes.[13] The process of gametogenesis differs in male and female mammals (Figure 2.12). Once female germ cells, which are generally more resistant to mutagens than male gametes, reach the oocyte stage in sexually mature mammals, DNA replication is suspended. The continuous processes of DNA synthesis and replication occurring in males probably account for the enhanced susceptibility of their germ cells to a wide range of genotoxic agents. Unfortunately, there are only minimal data on mutagenicity in female oocytes with which to test these assumptions.

## Cell Type

A critical issue that must be addressed in any analysis of genetic risk is the importance of somatic cell risk versus germ cell risk. Some mutagens may pose a risk to accessible somatic cells, but biological/physical barriers prevent these mutagens from reaching the germ cells in either the male or female. Although mutagenic damage to somatic cells appears to be a component of cancer risk, additional mechanisms such as tumor promotion must be factored into somatic cell risk estimates since malignant cell transformation appears to be a multistep process. Somatic analysis therefore becomes important in developing hazard assessments for cancer and teratogenesis and are integral dosimetry components of the parallelogram approach to germ cell risk.[14]

Gonadal tissues as well as the gametic cells residing within these organs appear to be protected from chemical exposure by a blood–gonad barrier and by physical barriers such as the membranes enveloping the ovary. Consequently, the germ cell dose of a chemical may often be less than the somatic cell dose following a given whole-body exposure. In fact, special terminology has been developed to describe the differences between applied and realized doses:

*Exposure dose*. Amount potentially available to enter organism; usually expressed as a concentration × time (dose rate).

*Pharmacologic dose*. Amount actually entering the organism; usually expressed as amount/body weight.

*Tissue dose*. Amount that reaches specific target tissue; may be either greater or less than the pharmacologic dose on an amount/body weight basis.

*Target dose*. Amount that reaches the nucleus and/or DNA-containing region of the cell.

*Molecular dose*. Amount that is associated with critical sites on the

DNA molecule; current attempts to define this dose involve the detecting of specific alkylation sites on DNA.

*Collective population dose*. The average exposure dose per individual multiplied by the number of individuals exposed; conversely, it can be calculated by dividing the total dose released into an environment by the number of individuals potentially exposed.

*Genetically significant dose*. Dose received by germ cells that can potentially affect the next generation; any dose received by individuals past reproductive age must be subtracted from the genetically significant dose.

One assumes that for many classes of chemicals the distribution of the biologically active moiety will be uniform throughout the organism, resulting in equal dose to both somatic and germ cells; in a few situations, active concentration of the agent in the gonads is possible. Thus, demonstration of, or, conversely, a lack of genetic risk to somatic cells may not be sufficient to predict or preclude risk to germ cells.

There is a tendency to assume that for a given mutagen somatic cell risk is greater than or equal to germ cell risk and that studies of tumor induction would be adequate to set the more sensitive risk for a chemical. While this presumption is often correct, studies of ethylene oxide, ethylene dibromide, and ethylnitrosourea indicate that these agents are potent mutagens to male germ cells[15] at low doses and that mutation may be of greater concern than neoplasia.

A complication limited to female risk assessment is that in the female, oocytes can be continuously exposed to environmental mutagens and may accumulate damage, whereas in the male, spermatozoa are continuously replaced and altered sperm may disappear after a fixed period following acute mutagen exposure. The potential for exposure accumulation in females may offset the inherent sensitivity of the maturing sperm in males.

## Diet

Superoxide and hydroxyl radicals are probably formed *in vivo* by the oxidation of lipids (e.g., fats during cooking).[16] These radicals, identified as lipid hydroperoxides and lipid expoxides, can interact with DNA, producing mutational lesions. These radicals vary with an individual's lifestyle (diet) and age.

Dietary constituents also affect risk, since some compounds act as enzyme inducers and modulate metabolic processes, or agents such as BHT and β-carotene can act as antioxidants and radical traps.[17] Many dietary components are themselves genotoxic or carry genotoxic residues from pesticides and therapeutic agents and may contribute to the genetic

TABLE 5.3
Summary of Factors Influencing Risk Analyses

| Contributing factor | Parameter affected | Association with mutagenic risk |
|---|---|---|
| Genotype | (1) Variation in metabolic capacities or specificities | (1) Toxification and detoxification increase or reduce systemic dose following fixed exposure |
| | (2) Variation in DNA repair capacities or specifications | (2) Decreased DNA repair capacities would likely reduce the biologically effective dose of a mutagen |
| | (3) Sex chromosomes | (3) Differences in turnover rate of germ cells in males and females influence sensitivity of the germ cells to a fixed exposure level of a mutagen |
| Diet | (1) Enzyme inducers or inhibitors | (1) See (1) above |
| | (2) Antioxidants (BHT, β-carotene, etc.) | (2) Many antioxidants bind free radicals, which are known to be genetically active |

risk of other chemicals via additive and/or synergistic responses. Table 5.3 summarizes the secondary factors that influence risk estimates.

## Theoretical and Practical Considerations Important to Risk Analysis

Several approaches to a unified concept of genetic risk have been proposed. The most notable and controversial is that of Abrahamson *et al.*[18] The hypothesis, designated ABCW, held that mutation rates per cell were a function of DNA content per haploid genome; and, when plotted, the rates showed a linear relationship across several different phylogenetic levels. The implications of this theory are significant, since it presumes a relatively simple extrapolation from microbes to man. This approach has been criticized because it does not take into consideration such parameters as the high levels of repetitive DNA normally found in mammalian cells, differential repair between species, and variable germ-cell-stage sensitivity.[19] The controversy has not been resolved, but it appears that is concept probably cannot be applied to all chemicals. The induction of genetic alterations is the expression of a series of complex interactions occurring in the target species. Some of the more important

TABLE 5.4
Comparison of Germ Cell and Somatic Cell Effects for Chemical Mutagens in Mammals for Commonly Used *in Vivo* Assays

| Compound detected as a germ cell mutagen | Assay[a,b] | Compound also detected in a somatic cell assay[b,c] |
|---|---|---|
| Ethylmethane sulfonate | G1, G2, G4 | S1, S2, S3 |
| Methylmethane sulfonate | G1, G2, G4 | S3 |
| Triethylene melamine | G1, G2, G3, G4 | S1, S2, S3 |
| Cyclophosphamide | G1, G2, G4 | S1, S2, S3 |
| Thio-tepa | G1, G3, G4 | S1 |
| Mitomycin C | G1, G3, G4 | S1, S3 |
| Myleran | G1, G4 | S1 |
| Colchicine | G1 | S2 |
| Trenimon | G1 | S1, S2 |
| Procarbazine HCl | G1, G2, G3 | S1, S2, S3 |
| Nitrogen mustard | G2 | S1, S2 |
| Ethylene oxide | G1 | S1, S2 |
| Ethylnitrosourea | G1, G3 | S3 |

[a] Germ cell assays: G1, dominant lethal assay; G2, HTA; G3, specific-locus assay; G4, spermhead abnormalities.
[b] Designations do not mean that unlisted assays were negative, but only that data were not available at this time.
[c] Somatic cell assays: S1, bone marrow cytogenetics; S2, micronucleus assay; S3, recessive mutation bone spot assay in mice.

factors include route of exposure, metabolic differences, and, in the case of germ cells, differences in blood–gonad barriers.

In the final analysis, risk also must be evaluated against societal or practical considerations, such as risk–benefit, risk–risk, and economic impact. The National Academy of Sciences defined this as risk management.[20] There will be unavoidable risks from naturally occurring mutagens or from materials impractical to remove from the environment. In these cases, it behooves society to know the relative, if not precise, impact of the risk. For example, it has been estimated that each human consumes about 10 tons of food (dry weight) by the age of 50, and some of this food contains naturally occurring or processing-derived mutagens.[16,21] There has been considerable speculation regarding the impact of these agents on our current genetic and cancer burden.[22]

Traditionally, estimates of genetic risk have been derived from studies of mutations in male and female gametes, since these cells provide the substrate for future generations. More recently, however, attention to somatic cell risk has increased due to improved technology and the realization that somatic cell damage significantly add to the disease load, specifically as causative factors for malignancy, terata, and heart disease (Chapter 3). *In vivo* somatic cell studies also act as surrogates for germ

TABLE 5.5
A Comparison of Risk Assessment for Somatic and Germ Cells *in Vitro*

| Comparison parameter | Somatic cells | Germ cells |
|---|---|---|
| Selective pressure on new mutants | Slight, since cells encounter only mitotic division | High, since meiotic cell division acts to screen out specific types of mutations, such as chromosome effects |
| Barriers to mutagenic agents | Few barriers if agent is distributed systemically | Barriers similar to those for somatic cells plus additional blood–gonadal barriers |
| Sex differences | Minimal; some influence of hormones on metabolism in specific organs | Significant because of the different processes involved in gametogenesis between male and female mammals |
| Ease of detection of induced effects | Each dosed animal represents a population of exposed cells | Analysis generally requires large animal populations, since each offspring represents a single exposed gamete analyzed |
| Size of exposed sample that can be scored | Extremely large population, generally $10^5$–$10^7$ surviving cells scored | Generally small sample sizes; fewer than $10^5$ animals in all cases |
| Type of genetic lesion detected | Chromosome aberrations and gene mutations both easily detected | Chromosome aberrations typically scored; point mutations require extensive studies to detect with reliability |

cell risk analysis, since the estimate of risk to somatic cells would most likely be on the conservative side. That is, a lack of mutation induction in somatic cells *in vivo* should indicate a corresponding lack of gamete risk. Examples supporting this assumption are shown in Table 5.4. Table 5.5 compares the variables involved in risk analysis between somatic and germ cells.

## RISK ESTIMATION METHOD FOR SOMATIC AND GERM CELLS

### Somatic Cell Risk

Four methods have been described for somatic cell risk estimation; three involve the use of animal models for direct measurement, and the fourth is an indirect approach combining *in vitro* cell culture with *in vivo* metabolism and chemical disposition. The first direct approach consists in the measurement of chromosome aberrations in bone marrow or other somatic cells of the exposed organism. These studies can be performed in animal model experiments or by direct evaluation of human populations through peripheral blood lymphocyte analysis. Problems abound, however, when an attempt is made to interpret risk on the basis of chromosome breaks, fragments, and other aberrations that are usually lethal to the affected cell. Risk estimation must be based on the level of stable, heritable aberrations observed. Therefore, analysis for sentinel aberrations known to be essential in the formation of stable aberrations may be of some value. One such approach is to measure the frequency of dicentric (double centromere) chromosome aberrations. Dicentrics occur infrequently in mammals (~1/2000 metaphases) and are related to the formation of heritable translocation aberrations. This low spontaneous rate necessitates rather large experiments to detect chemically induced aberrations, but it is a feasible approach since slight changes can be measured. Measuring chromosome aberrations in bone marrow or other somatic cells is not a measure of gene mutations and cannot be considered a conservative method for risk evaluation if used alone, because the relationship between concentration and type of mutation induced favors gene mutations at low doses.[23] An argument in favor of the cytogenetic analysis for risk estimation, however, is that over half of the human genetic disease burden (number of affected newborns) is due to chromosome abnormalities (Table 5.1).

The second direct approach was developed by Dean and Senner in 1977 and is a method of mutation detection in Chinese hamsters.[24,25] This method consists in exposing Chinese hamsters to mutagens, recovering tissues from the treated animals, preparing primary cell suspensions, and measuring the primary cells for forward mutation to 8-azaguanine or ouabain resistance induced in the DNA while cells resided in the animals. This method measures gene mutation induction and appears more sensitive than cytogenetic analysis, although the experience with this technique is limited. Presumably, this method can be extended to an analysis for target tissue susceptibility; Dean and Senner have detected mutation induction in cells derived from the lung, liver, bladder, kidney, and stomach of Chinese hamsters.[24,25] This technique is a direct assessment of mutation and requires only a small number of animals. Numerous technical difficulties involved with this technique remain to be solved; probably the most critical is developing the ability to increase viability and cloning efficiency of trypsinized primary cells. Application of this technique to human peripheral lymphocytes has been reported by Strauss and

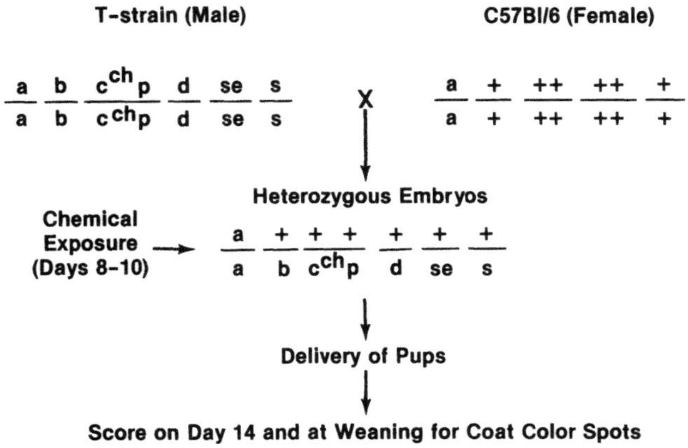


FIGURE 5.1. Description of the basic protocol for measuring the induction of somatic mutations *in vivo*. Specific inbred strains are employed in this test. The target cells are the embryonic melanocytes carrying heterozygous alleles controlling coat color. The pups are exposed *in utero*, and their hair is examined for spots (see Figure 8.4). Alleles: a, agouti; b, brown; $c^{ch}$, chinchilla; p, pink-eyed dilution; d, dilute; se, short ear; s, spotted.

Albertini.[26] These investigators demonstrated increased mutation to 6-thioguanine resistance in humans exposed to chemotherapeutic agents.

The third *in vivo* approach to direct somatic cell analysis involves the measurement of coat color spots in $F_1$ mouse pups. The basic method (Figure 5.1), originally described by Russell and Major in 1957,[27] consists in exposing pregnant female C57BL mice (mated to T males) near day $10\frac{1}{2}$ postconception.[3,15] The number of target cells (melanocyte precursor cells) per pup at this point is between 150 and 200. If it is assumed that the minimum number of scorable loci affecting coat color is four, then the mean number of cells at risk per litter of five pups is approximately 3500. If 50 pregnant females per dose group are used, the number of target cells at risk is 175,000. This represents a sizable number for data analysis, and a report by Russell indicates that the somatic test is more sensitive than the associated specific-locus germ cell test.[28] Comparisons with seven mutagens showed that the dose required to demonstrate mutagenic activity was lower for melanocyte precursors (somatic test) than for spermatogonia (germ cell test); with three of the mutagens, hycanthone, benzo[*a*]pyrene, and dimethylnitrosamine, no mutants were detected in spermatogonial experiments (Table 5.6). Russell has attributed the production of coat color spots to several types of somatic lesions, including point mutation, small chromosomal deletions, somatic recombination, and whole chromosome loss.[3] A growing data base has been accumulated for the somatic mutation coat color assay. A summary of the studies

TABLE 5.6
A Comparison of the Sensitivity of Somatic and Germ Cells to Specific-Locus Mutation Induction by Ten Agents[a]

| Chemical | Gene mutation response | | Activity ratio[d] |
|---|---|---|---|
| | In germ cells[b] | In somatic cells[c] | |
| Ethylmethane sulfonate | + | + | >1 |
| Methylmethane sulfonate | + | − | >1 |
| Triethylene melamine | + | + | 2.8 |
| Mitomycin C | + | + | 2.6 |
| Benzo[a]pyrene | +? | + | 5.0 |
| Hycanthone | − | + | >1 |
| Diethylnitrosamine | − | + | >1 |
| Natulan | + | + | 4.0 |
| 2-napthylamine | − | + | >1 |
| *N*-Methyl-*N*-nitrosoguanidine | − | + | >2.0 |
| Ethylnitrosourea | + | + | 3.0 |

[a] Modified from Russell[3] and Styles and Penman.[29]
[b] Effect based on data from spermatogonia or spermatozoa (whichever appeared more sensitive).
[c] Data from the mouse *in vivo* somatic coat color mutation assay.
[d] Effective germ cell dose/effective somatic cell dose = activity ratio. Values >1 suggest greater sensitivity in the somatic cell test.

reported with the somatic cell assay has been given by Styles and Penman.[29]

The target cell numbers associated with this assay are reasonably large, and the test does not suffer from the technical limitations described for the tissue culture analysis of Dean and Senner in Chinese hamsters. However, differential placental transport of chemicals among species has been well documented,[30] and uncertainty of adequate fetal exposure complicates interpretation of negative data from the somatic spot test.

The fourth approach is an indirect method of determining somatic cell risk. This integrated technique couples *in vitro* mutation induction with DNA adduct formation. Alkylating agents as well as other reactive molecules form covalently bound DNA adducts at various sites that can be measured biochemically if the chemical is tagged with a radiolabel such as $^{14}C$. Other techniques, using immunofluorescence detection systems, are very sensitive and can detect a single adduct per $10^6$ nucleotides. Using these methods to measure the dose, one can define the activity of an agent as a ratio of

$$\frac{\text{induced mutants}}{\text{units of alkylated DNA bases}} = \text{mutagenicity index (MI)}$$

Standard MI curves for specific reference chemicals of different adduct-forming classes can be developed for use in evaluating "unknown" samples. By dosing animals with the labeled "unknown" test substance, collecting somatic tissue, and calculating the number of DNA adducts produced per genome, one can calculate the projected level of induced mutants per cell from a standard mutation/adduct curve. This approach can be performed rapidly and with a high degree of sensitivity assuming that the test material is amenable to radio- or fluorescent-labeling and forms DNA adducts.

Cells from various suspect target tissues can be analyzed. One important limitation of the DNA adduct method is that different alkylation or adduct sites appear to be associated with different levels of induced mutations. For example, alkylations at the O-6 position of guanine or the O-4 position of thymine are more likely to result in mutation than alkylation of the N-7 position.[31] This type of information will be critical for quantitation of effects and must be employed in determining the potential genotoxic risk for a given agent.

A review of these and other methods for *in vivo* somatic and germ cell mutagenesis has been published by Malling and Valcovic.[32]

## Germ Cell Risk

Three important factors must be assessed in an evaluation of genetic risk.

1. The DNA in male germ cells passes through a series of replication and meiotic steps that exhibit differential sensitivity to chemicals. Cell-stage sensitivity is important in risk estimates, since some cell stages are of greater significance to risk values than are other stages. For example, mutagenesis studies of chemicals in pre- and postmeiotic germ cells of mice show that premeiotic cell stages are highly refractory to induced mutation compared with spermatozoa (Table 5.7).

2. There are clear sex differences in sensitivity to chemical mutagens.

3. Mutations induced in one generation are spread across several generations because they persist (if not lethal) and can be linearly passed along through several generations. Therefore, new mutations in each subsequent generation must be added to those first induced in the previous generation. The only mutations not persistent are dominant mutations that produce lethality or sterility in the affected individual. The length of time a new mutation will persist in the human population is generally inversely proportional to its severity.

## Indirect Measurement Approach to Risk Estimation

The expression of genetic risk in humans is normally expressed in one of two ways (see ICPEMC[33] for a detailed explanation). The first is

TABLE 5.7
Relative Sensitivity of Germ-Cell Stages of Male Mice to Dominant Lethal Induction by Chemical Mutagens[a]

| Compound | Sensitivity at given germ-cell stage | | | |
|---|---|---|---|---|
| | Spermatozoa | Spermatids | Spermatocytes | Spermatogonia[b] |
| Fosfestrol | + | − | − | − |
| Mitomen | + + | − | − | − |
| Ethylmethane sulfonate | + + | + + | − | − |
| Methylmethane sulfonate | + + | + + | − | − |
| Propylmethanesulfonate | + + | + + | − | − |
| Cyclophosphamide | + + | + + | − | − |
| Isopropylmethanesulfonate | + + | + | + + | (−) |
| Myleran | + + | + + | + + | (−) |
| Natulan | − | + | + + | (−) |
| Mitomycin C | − | + | + | (−) |

[a] From ICPEMC.[33]
[b] Parentheses indicate failure of fertilization rather than dominant lethal induction.

the proportional or indirect method. This method ensures that a given rise in mutation rate in humans will result in an equivalent rise in genetically based disorders. The most common unit of expression is the dose required to double the spontaneous mutation rate (i.e., the doubling dose). This value is based on linear extrapolation of dose–response data from animal studies and is expressed as a ratio of induced mutation rate per dose to the spontaneous rate. It is necessary to have an accurate estimate of the background mutation rate.

Underlying assumptions for the doubling-dose calculation are linearity of the dose–response relationship and sensitivity equivalence between mice and humans. The analysis is based on the assumption that the gene (or genes) assayed in the animal model serves as surrogate(s) for all other genes and that all genetic disease will be increased proportionately. Table 5.8 shows examples of the doubling doses for humans calculated from mouse specific-locus test data.

## Direct Measurement Approach to Risk Estimation

The second method is direct calculation of an increased mutation rate per unit dose in absolute terms and the resulting number of new cases of disorders caused by a specific genetic mechanism induced. The spontaneous rate for specific effect does not need to be known. Ideally, one would like to use human data for these calculations; however, the most

TABLE 5.8
Comparison of the Doubling Doses for Six Chemotherapeutic Drugs Based upon the Results from the Mouse Specific-Locus Assay[a]

| Compound | Experimental dose used (mg/kg) | Calculated doubling dose (mg/kg) | |
|---|---|---|---|
| | | Spermatogenic | Postspermatogenic |
| Procarbazine | 600 | ~98 | 43 |
| Mitomycin C | 5.25 | 1 | NA |
| Isoniazid | 2 × 125 | 219 | NA |
| Triethylene melamine | 2 × 2 | 0.3 | NA |
| Busulfan | 20 | 6 | NA |
| Hycanthone | 150 | 60 | NA |
| Cyclophosphamide | 120 | 320 | 4 |

[a] Adapted from Ehling.[34,35] NA, not available.

data used to calculate genetic risk comes from animal experiments, such as the mouse specific-locus assay, mouse heritable translocation assay, and dominant viable mutation assays in mice. Table 5.9 provides a list of assays used in genetic risk analysis with relevant literature citations. The relationship between results in short-term tests and risk assessment tests has been reviewed by Gene-Tox.[41]

The concept of heritability of germ cell effects can be a confusing issue, in that many of the available *in vivo* techniques do not define true heritability. The dominant lethal test assesses damage to male gametes that is transmitted as far as the embryo. Lethality is the endpoint and heritability of the damage cannot be determined. A similar situation exists for the heritable translocation assays. The chromosome alterations in this assay are transmitted to $F_1$ progeny, but result in sterility or semisterility, which is not a heritable trait. Consequently, neither of these two assays are true measures of the multigenerational heritability of induced DNA damage. Specific-locus mutation tests in mice do result in effects that are germ cell transmissible and heritable over several generations. Utility of the spermhead morphology assays is less well defined because the mechanism and heritability of spermhead abnormalities are not completely elucidated.

Direct approach calculations also assume absolute equivalence in dosimetry and response factors for rodents (mice) and humans.

## Alternative Indirect Methods

Sobels[14] proposed a ratio method identified as the parallelogram approach. This approach also assumes equivalence between experimental

TABLE 5.9
Animal Methods Employed for Genetic Risk Analyses

| Bioassay | Technique | Reference |
|---|---|---|
| Specific-locus test in mice | Measurement of visible autosomal recessive mutation in mice using a multiple marked tester strain of mice containing up to seven recessive loci; minimal size of experiment needed to define negative would be in the range of 12,000–14,000 offspring of treated animals | 36 |
| Cataract mutation assay in mice | Measurement of visible dominant mutations in mice using a biomicroscopic examination of the eye with a slit lamp; $F_1$ offspring of treated mice are examined at 3 weeks of age; no special strains of mice are required; this assay can be used to make direct calculations of induced mutation rates | 35, 37 |
| Skeletal mutation assay in mice | Measurement of visible dominant mutations affecting the skeletal system of mice; requires careful analysis of bone development | 38 |
| Heritable translocation assay in mice | Measurement of the incidence of balanced translocations transmitted to the $F_1$ progeny from treated males; the phenotype is detected by observation of sterility or semisterility in the male $F_1$ offspring and confirmed by cytogenetic analysis | 39, 40 |

animals and humans. The basic logic is that of the ratios of $B/A$ and $C/D$ are constant,

$$B/A = C/D$$

and values for $A$, $B$, and $D$ can be measured, the value for $C$ can be calculated.

As an example, let the relationship between ($A$) the number of chemical–DNA adducts formed per unit dose in lymphocytes and ($B$) the number of specific-locus mutations in the male germ cells be expressed as a ratio. The relationship of somatic/germ cell events is then established for rodents with that agent. The next step is to measure the number of chemical–DNA adducts ($D$) formed in the lymphocytes of humans exposed to the test chemical. The specific-locus mutation rate for a given exposure ($C$) is then estimated using the assumption that the ratio of adducts to mutation in rodents is predictive of the ratio in humans.

The parallelogram approach is advantageous because it utilizes data from humans, and does not require an accurate measurement of test-compound exposure. The second indirect method assesses somatic-cell risk for cancer initiation using an index developed between cancer rates and *in vitro* test data potency for known human carcinogens.[42] This ratio is used in a similar fashion to Sobel's parallelogram to calculate an index for an unknown agent using *in vitro* test data in combination with the risk ratio.

## DOSIMETRY ISSUES IN RISK ASSESSMENT

### Relationship of the Route of Exposure to Total Body Dose

Genetic toxicology studies have been conducted and reported using all routes of compound administration. Table 5.10 lists the most common methods of exposure and their limitations. The total dose received and its distribution varies considerably according to the route selected. For example, rats exposed to 500 ppm of a compound will be subjected to different total real doses according to the route of administration.

Using current toxicologic procedures, it is virtually impossible to precisely calculate total body dose for chemicals encountered under normal conditions. However, for the purpose of estimating risk, one can rank the relative risk associated with certain types of exposures. Table 5.11

TABLE 5.10
Routes of Exposure and Their Limitations

| Routes | Limitations |
|---|---|
| Oral<br>Feed<br>Drinking water<br>Gastric intubation | Must identify stability in feed or water as well as intake during exposure; places stress on animals and gives agent by pulse rather than continuous dose |
| Inhalation<br>Respiratory<br>Intratracheal insufflation<br>Implantation | Inhalation dose monitoring is technically difficult for certain substances, especially complex mixtures of many different substances; there can be a significant amount of ingestion of material carried in the hair during grooming after exposure; particle size is important in defining the actual total-body dose |
| Dermal<br>Skin painting<br>Injection (subcutaneous)<br>Intravenous<br>Intraperitoneal | Localized immunologic or irritation effects can cause problems; skin absorption must be identified to understand dose |

TABLE 5.11
The Relative Risk Resulting from Exposure to a Model Compound[a]

| Route of exposure[b] | Relative level of risk[c] |
|---|---|
| Inhalation (nonparticulate) | Highest |
| Diet (feed) | ↓ |
| Drinking water | |
| Dermal (skin contact) | Lowest |

[a] The model compound produces optimal absorption under all conditions and is not metabolically altered.
[b] Total dose is expressed as concentration × time (e.g., 500 ppm for 24 hr).
[c] Based on a total-body dose achieved following exposure to the specified concentration by various routes.

shows the ranking based on the volume of a compound likely to be consumed and/or absorbed during a given period of time at a fixed concentration. While the relationships shown in the table were developed from rodent data, they apply to all mammalian species. This aspect of dosimetry is critical to risk methods not using a direct measurement of reaction products at the target molecule.

A similar level of uncertainty exists for target organ dose, since the route of administration affects the type of primary metabolism of the sample. The microsomal P-450 enzyme composition of lung tissue may differ from that of liver tissue and may assume the production of qualitatively or quantitatively different metabolites at the primary metabolic site. This difference could influence the target organ effect prior to liver activation.[43] Thus, route of exposure and portal of entry are important considerations in analyzing chemical fate *in vivo*.

## Relationship of the Route of Exposure to Metabolism

The *in vivo* distribution and metabolic fate of a chemical are dependent to a certain extent on the route of exposure. For instance, oral dosing has the advantage of allowing for a dose of significant mass, but it can give misleading information if the test substance is acid-labile, resistant to gut transport, susceptible to bacterial degradation, or becomes readily trapped by the excreta. As a result of oral administration, the test substance enters the hepatic portal system and is, of course, immediately subjected to the liver and its profound metabolic activity. Once leaving the hepatic circulation, the test substance is then exposed to the general circulation.

Intravenous dosing, on the other hand, has the advantage of direct introduction into the general circulation. A large volume of the test substance is usually not possible via intravenous injection because of high direct toxic effects of the material on the experimental animal. Furthermore, in the case of lipophilic products, a suitable vehicle must be found, and this further complicates the dosing situation. Once these problems are overcome, however, the product is in the general circulation, and the most elucidative picture of specific organ metabolism and toxicity is usually revealed. Intraperitoneal injections are not that dissimilar to intravenous injection, since the test substance quickly reaches the general circulation and gives the same pertinent information.

Dermal application or dosing has numerous problems, involving dermal trapping, evaporation, physical removal (e.g., licking or grooming), and poor absorption. There is known metabolic activity in the skin, and the pharmacologic events may begin immediately upon application. It is probably fair to presume, however, that once transported into the general circulation, the test substance proceeds as if given intravenously.

Inhalation dosing and the subsequent metabolic studies would probably have to be well controlled so that dermal exposure does not occur simultaneously. The lung is a proven metabolic organ, and this could provide initial alterations in the test substance. Here again, following transport across the lung and its initial metabolic impact, the test substance would be in the peripheral circulation for general metabolism.

## Blood–Gonadal Barriers

Levels of chemicals in the bloodstream in the systemic circulation system can be used to approximate somatic cell exposure, but are not necessarily accurate for germ cell exposure, since the gonads are somewhat protected from the general circulation by what are referred to as blood–gonadal barriers. The movement of nonelectrolytes from the body circulation to the testes, for example, is dependent on molecular size.[44] Molecules smaller than 3.6 Å can be transported readily, while larger molecules cannot. Lipid solubility is also important, and studies suggest that this is a strong rate-limiting factor in penetration at physiologic pH.[44] There is minimal evidence indicating active transport in the testes, although certain antibiotics and metal ions appear to be concentrated in rete testis fluid to a greater extent than they are found free in plasma.

Penetration of chemicals into the ovary also has been studied.[30] The ovarian follicle is an avascular structure, and before a chemical can reach an oocyte located in a maturing follicle it must leave the vascular system, cross the thecal layer, and enter the follicular fluid.[30] Once in the follicular fluid, an agent moves passively into the oocyte. The same factors that affect transport across the blood–testis barrier (molecular size, lipid solubility, and ionization) also affect follicular transport in the ovary.

In general, then, gonadal structures restrict the permeability of many foreign compounds to the mature male and female gametes (spermatogonia, however, may lie outside the blood–testicular barrier). Little active or selective transport occurs. The major transport of chemicals depends on their lipid solubility and molecular size, and penetration appears to obey simple diffusion kinetics. As a result of this restriction of transport, the gonadal barriers probably reduce the risk of exposure of germ cells compared to somatic cells.

## Molecular Dosimetry

Dose is generally described as a function of concentration and time. However, this approach to dose calculation does not take into consideration several problems, such as the nonlinearity of exposure to dose ratios, rapid loss or alteration of the molecule during exposure, and differences in metabolism as a function of concentration and length of exposure. The most accurate measure of dose is the quantification of reaction products as described in the section on somatic cell risk. Using alkylating agents as models, one can define alkylated DNA bases as the unit of dose determination. This approach may not be useful for studies of certain mutagens, such as intercalating mutagens, which do not covalently bind to macromolecules. Lee[45] reported the relationship between EMS dose in *Drosophila* and mice and ethylations per sperm cell to be linear. Data from SLRL studies in *Drosophila* showed that a linear relationship also existed between ethylations/nucleotide and percent mutation.[45] If mutation per locus per alkylation is roughly equivalent across species, then extrapolations across mammalian species should be possible. The ability to perform this type of dosimetry is almost unique to genetic toxicology, where the specific target molecule is known and can be examined with very sensitive methods. Figure 5.2 shows the ethylations/nucleotide relationship for recessive lethal mutation, dominant lethal mutations, and reciprocal translocations as derived by Lee.[45]

Two additional considerations are important in risk calculations. One is the effect of DNA repair processes on adducts or other lesions induced (see Chapter 6). It has been demonstrated that the efficiency of repair is dependent on the type of reaction product formed and the meiotic cell stage exposed. Many of the metabolic and repair capabilities are suppressed in mature gametes, making them somewhat variable in sensitivity, depending on the nature of the chemical. Direct-acting alkylating agents, for example, are very effective mutagens for mature sperm, but not so effective for premeiotic spermatogonia.[46,47] There are also quantitative

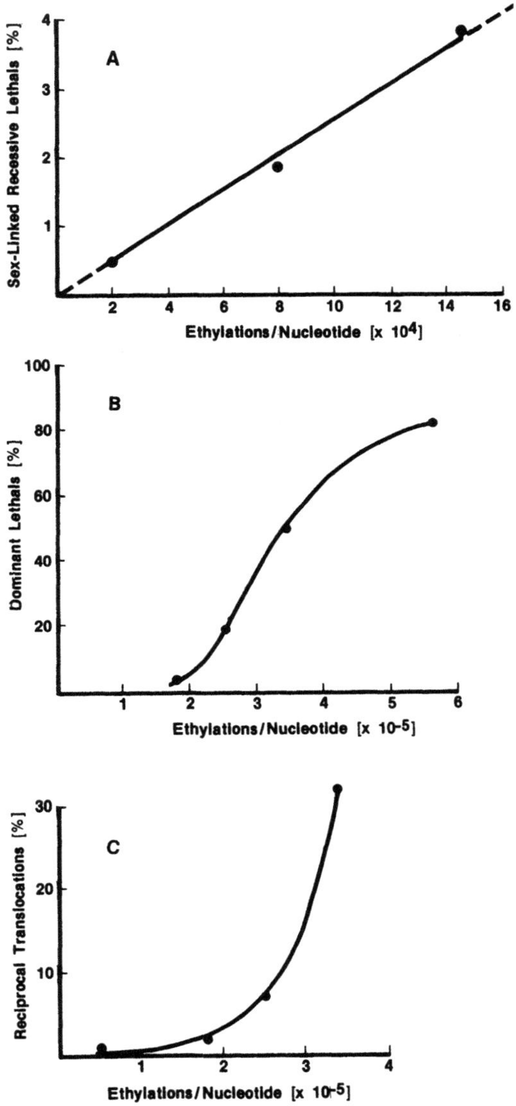


FIGURE 5.2. Relationship between the formation of reaction products and genetic effects in *Drosophila*. (A) Dose as alkylations per nucleotide versus induced frequency of sex-linked recessive lethals in spermatozoa following 24-hr feeding of EMS to adult *Drosophila melanogaster*. (B) Dose as alkylations per nucleotide versus dominant lethal frequency *DL* computed by the following formula: *DL* = 1-living embryos as a percentage of control. Sperm were sampled 6.5–9.5 days after injection of the male, and only data from fertile matings were used. (C) Dose as alkylations per nucleotide versus relative frequency of reciprocal translocations. Both sterile and partially sterile $F_1$ males were used.

differences in repair capacity between species, which can affect the number and types of adducts remaining in exposed DNA. Another factor is target size. Eukaryotic organisms, mammals in particular, have a high content of repetitive DNA (nonfunctioning DNA) in their chromosomes.[32] Since the amount of DNA appears to influence the yield of mutation, dosimetry must be able to define an effective dose (i.e., number of reaction products per active gene) taking into account the amount of repetitive DNA sequences in a given species.[48]

## GENE FUNCTION ISSUES IN RISK ASSESSMENT

It has been estimated that the haploid human genome contains about $6 \times 10^9$ nucleotides, or some 30,000–50,000 genes.[32] Genes play different roles in cellular activities, and since the manner of expression of an altered gene depends on the role of the gene in the organism, the selection of a gene locus for risk analysis may affect the outcome. Table 5.12 summarizes both the kinds of functions controlled by genes and the probability of directly detecting a point mutation induced in such a gene. Detection of altered structural genes is the usual method of mutation identification because the mutant phenotype can usually be readily identified (e.g., the

TABLE 5.12
Location of Genetic Effects and Ability to Detect Effects in the Target Molecule[a]

| Type of gene | Function | Ease of direct detection |
|---|---|---|
| Structural gene | Codes for the amino acid sequence in a protein (Figure 2.2) | Easily detected—forms the basis of most genetic assays used in routine screening |
| Operational genes | | |
| Modifying gene | Modifies the protein structure after the primary polypeptide forms | Can be measured by genetic tests, such as those measuring temperature-sensitive mutations; however, not commonly detected in routine screening tests |
| Architectural gene | Determines the organ in which enzyme expression will occur | Extremely difficult to identify; almost never detected in screening |
| Temporal gene | Determines the time in development at which a structural gene is turned on or off | Extremely difficult to detect; almost never detected in screening |
| Regulatory gene | Modifies the production of gene products in response to internal or external environments | Can be detected in microbial systems, but is seldom used in screening |

[a] Modified from Malling and Valcovic.[32]

requirement for nutrients in bacteria or hair color changes in mice). Detection of operational genes usually requires highly specialized research methods. However, mutations occurring in nonstructural genes could contribute substantially to the induced genetic burden.

## DATA EXTRAPOLATION ISSUES IN RISK ANALYSIS

Using test data from all available sources, the genetic toxicologist must try to extrapolate the results of *in vitro* and rodent model tests to the human experience.

1. Most tests, especially *in vivo* ones, are conducted with small test populations at high exposure levels. The proper extrapolation of results down to the typical environmental exposure levels for large populations is disturbingly uncertain, and the estimate of human risk depends on the statistical model employed. The uncertainty of risk estimation at low dose levels is clearly demonstrated in Table 5.13, where the estimated increase in cancer incidence from exposure to saccharin was found to be from 0.0007 to 3640 cases per 50 million population per year, depending on the statistical model used.

2. There are likely to be substantial pharmacodynamic differences between species, and these can lead to errors when judging susceptibility. For example, thalidomide, the human teratogen, is considerably more

TABLE 5.13
Estimated Human Risks from Saccharin Ingestion at 0.12 g/day[a]

| Method of high- to low-dose extrapolation | Lifetime cases per million exposed | Cases per 50 million/year |
|---|---|---|
| Rat dose adjusted to human dose by surface area rule | | |
| Single-hit model | 1200 | 840 |
| Multistage model (with quadratic term) | 5 | 3.5 |
| Multihit model | 0.001 | 0.0007 |
| Mantel–Bryan probit model | 450 | 315 |
| Rat dose adjusted to human dose by mg/kg/day equivalence | | |
| Single-hit model | 210 | 147 |
| Multihit model | 0.001 | 0.0007 |
| Mantel–Bryan probit model | 21 | 14.7 |
| Rat dose adjusted to human dose by mg/kg lifetime equivalence | | |
| Single-hit model | 5200 | 3640 |
| Multihit model | 0.001 | 0.0007 |
| Mantel–Bryan probit model | 4200 | 2940 |

[a] From NRC/NAS Report on Saccharin.[49]

effective in humans than in the typical rodent model test system.[50] Thus, preliminary risk estimations using rodent species would have erred in the wrong direction.

3. Basic differences in physiology and organ pharmacodynamics may also play an important role in extrapolation of risk. For example, the intrinsic capacity of DNA repair systems is different in rodent species from that of humans and repair capacity can be extremely important when considering extrapolation of effects at usage levels where most lesions may be corrected. Pulmonary physiology, immunocompetence, placental permeability, renal function, and hormonal variability are also different, and as a result may directly influence the expression of an effect in various mammalian species.

4. The degree of potency must be considered in extrapolations of animal data. Potency is the intrinsic activity of a mutagen, and is a product of the chemical's DNA affinity and site specificity.

## Extrapolation of *in Vitro* Potency to *in Vivo* Response

As indicated in Chapter 2, numerous attempts have been made to establish a relationship between the mutagenic activity of carcinogens *in vitro* and their carcinogenic activity *in vivo* (typically in rodent bioassays). With the *Salmonella* system, Meselson and Russell[51] were able to demonstrate a rough relationship between their *Salmonella* data and carcinogenicity. Similar results have been reported from Ames' laboratory and a major effort to expand this data base is in progress. A second comparative report using the *in vitro* gene mutation system in L5178Y mouse lymphoma cells as the mutagenic parameter was conducted by Clive *et al.*[52] Their data also demonstrated a potency correlation between *in vitro* and *in vivo* responses.

These studies have raised hopes that categories of carcinogenic potency could be determined by *in vitro* mutagenesis assays. Critics quickly pointed out several potential problems with this approach.

1. Selection of the specific target cell for mutagenic analysis can greatly influence the reported potency of the chemicals being compared. For example, *Salmonella* mutants TA1538 and TA98 are both derived from 3052, yet for a given chemical mutagen significantly quantitative differences between strains may be obtained. No rules have been formulated to decide which should be used in developing the correlation.

2. If a target organism is selected that gives apparent correlation under one set of test conditions, the correlation may be significantly altered by slight modifications in the study design. This suggests that a multitude of different study designs would have to be used to determine

which gives the proper correlation, and adoption of uniform study designs probably would not yield the proper consistent relationships.

3. *In vitro* metabolic activation systems most often employ S9 homogenate preparations. These preparations are prepared from animals pretreated with inducing agents and do not accurately reflect normal *in vivo* metabolic processes. Choice of species, inducers, and levels of S9 employed in the *in vitro* assays can modify the degree of response *in vitro* and significantly modulate the correlation of a chemical within a given test system.[53] The use of intact hepatocytes, which probably give more accurate activation responses, will give yet a different set of correlations.

4. If data points from noncarcinogenic mutagens and mutagenic noncarcinogens are added to the comparisons shown in Figure 3.11, the picture becomes substantially more complex and less convincing.

Because of the ease with which *in vitro* potency can be modified via the methods described above, it is unlikely that an exact *in vivo* response can be predicted with accuracy. A general relationship for some types of chemicals (those that do not require activation) may be observed between *in vitro* and *in vivo* responses, but the only reliable methods available for assessing *in vivo* potency are with animal models.

## Extrapolation from Species to Species

Differences in metabolism and body distribution of chemicals vary widely among animal species. Examples are common of one species unable to convert a promutagen/procarcinogen to its active form while other (even closely related) species can. The following are additional examples of factors affecting the qualitative and quantitative responses across species lines.

1. The number of transcribed genes per unit length of the genome changes substantially as one proceeds up the phylogenetic scale. Almost 100% of the genome of a bacteria such as *E. coli* codes for structural or associated potential gene target. As species move up the phylogenetic scale toward humans, the number of functional (transcribed) genes per unit length of the genome is strikingly reduced. In humans, it has been estimated that only 1–2% of the entire genome codes for structural genes. Thus, the critical target size versus total mass of DNA appears to vary considerably among species.[32]

2. The mechanisms of mutation induction in different species appear to be different. In bacteria, specific inducible repair enzymes and mutation processing genes are directly involved in mutation induction.[54] Similar repair steps and mutation processing genes appear to be absent or are significantly more complex in eukaryotic cells. This raises questions con-

cerning how DNA lesions are processed in phylogenetically different organism.

3. Within placental mammals, DNA repair, replication rates, and expressibility appear to vary significantly within somatic cells both *in vivo* and *in vitro* as a function of cell type, organ, species, and age, and may be modified by various higher order physiologic functions, such as diet, stress, and hormonal levels.[55–57]

## Threshold

Discussions of risk estimation and extrapolation invariably lead to the subject of threshold levels for toxic effects. Certainly in the area of reversible toxicity, "no-recognizable-effect" levels can be established. The same cannot be confidently stated for irreversible effects such as mutations and oncogenic transformation. The types of animal bioassays required to establish threshold levels for mutation or tumor induction are prohibitive, and the debate thus becomes academic. However, for purposes of decision-making and many types of relative risk estimation it is worthwhile to consider the concept of "no-observable-effect levels" (NOEL). These values represent concentrations of the test substance where health effects, if they exist, are less than the normal background variation in effects in the general population.

Considering the available data, certain broad generalizations can be made regarding thresholds for mutation induction.

1. Gene mutation induction at high dose levels used in chemical screening protocols appears to be a one-hit phenomenon that extrapolates linearly to zero dose. This conclusion is derived from experiments conducted with agents at concentrations where the DNA repair system is has been saturated. Under these conditions the influence of repair on mutation induction kinetics is near zero. However, effects of DNA repair processes at the low end of the dose range where repair systems are not saturated may modify the shape of the dose–response curve. In fact, data from *in vitro* assays using normal cells and repair deficient target cells (surrogates for high-dose saturation) support nonlinearity of response at low concentrations (Figure 5.3).

2. Agents that modify the mutagenic activity of chemicals (comutagens) generally have no-effect levels. For example, mutagenicity studies combining initiating agents and promoting agents have demonstrated modulation of mutagenic responses. These agents act through biochemical triggers that are clearly concentration dependent, and low concentrations of these modulating agents are usually not active enhancers.

3. Chemicals that are specific for clastogenic effects via disruption

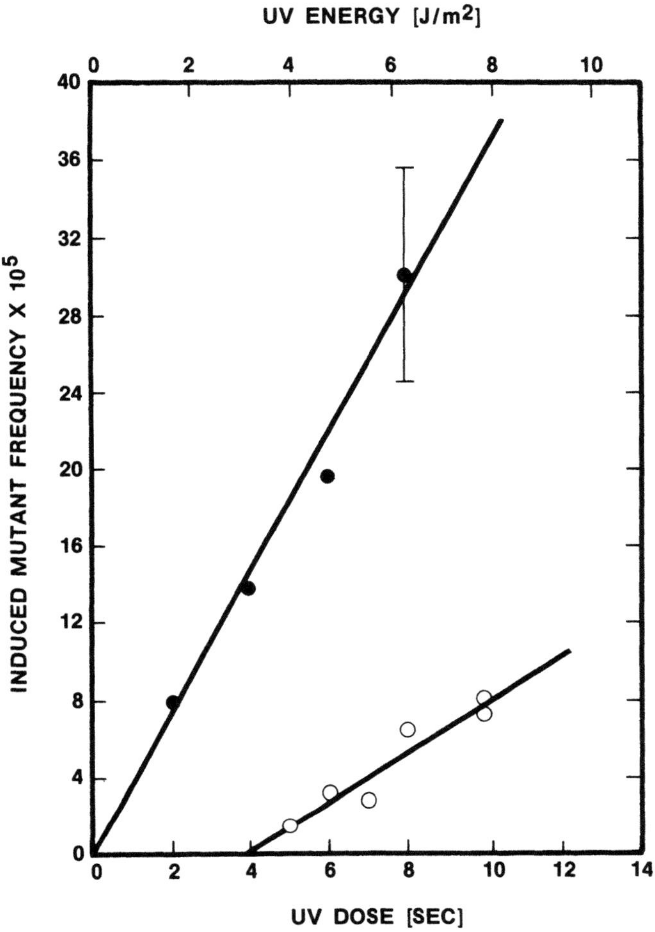


FIGURE 5.3. The effect of DNA repair on the mutation frequency in human fibroblasts. An apparent "no-effect" dose is observed in (○) normal and (●) XP4BE repair-deficient human diploid fibroblasts. The ability of a cell to repair DNA damage clearly shifts the slope of the dose–response curve. Data courtesy of Myhr *et al.*[58]

of spindle apparatus (e.g., colchicine and Benomyl) appear to have clear-cut thresholds.[59]

4. The approach used in interpretation of dose–response data may yield different conclusions regarding threshold responses. Figure 5.4 shows two possible interpretations of a set of epidemiologic data.

Assuming that oncogenic transformation is dependent on a genotoxic lesion, the factors previously described for mutagenic thresholds will also

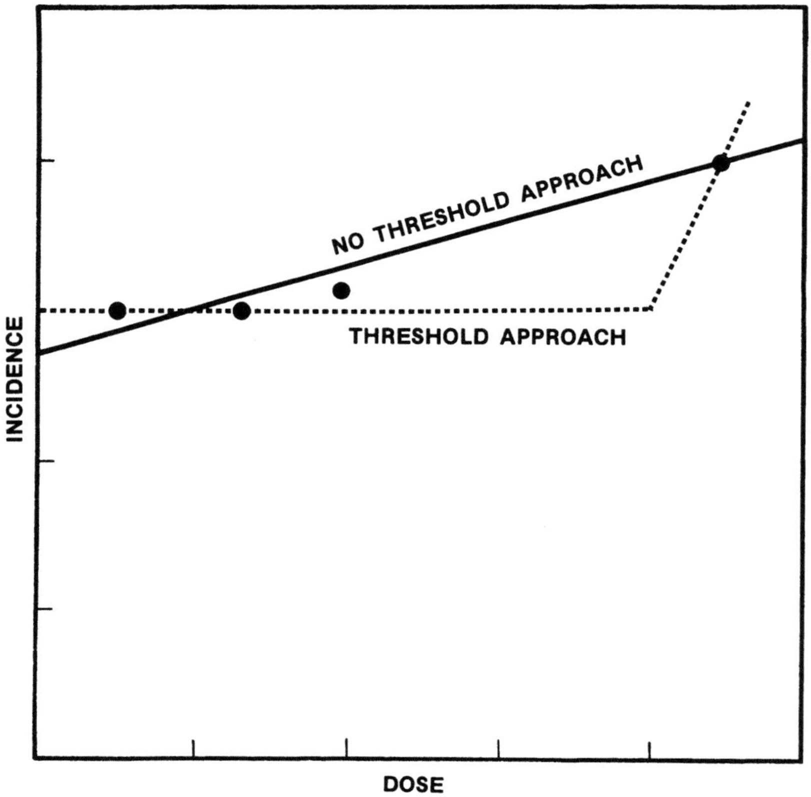


FIGURE 5.4. Two views of test data from a multidose study. The same set of data is shown graphed in two ways. Each method of data presentation supports a different interpretation.

apply for carcinogenic thresholds. Since it is unlikely that definitive animal studies will be conducted to prove nonlinearity of mutagenic responses, general agreement on risk-assessment models and levels of acceptable risk need to be defined by the scientific community.[60]

## IS GENETIC RISK IMPORTANT TO SOCIETY?

Unlike other types of chemically induced toxicity, the emotional burden and health-care costs associated with genetic alterations induced in the gonads of reproductively active humans will be borne by future generations. The present generation is already paying substantial amounts of money for the treatment and care of people affected by diseases of genetic origin (health care costs for genetic disease have been estimated to be at least 25% of the total dollars spent on medical care in the United States); it is the responsibility of our society at the very least to pass on a gene pool with a genetic burden no greater than it received.

Another issue is that of persistence. Genetic lesions persist in the human gene pool for varying periods of time, ranging from one generation (trisomies or some dominant mutations) to many generations (recessive specific-locus mutations). Consequently, increases in the genetic load caused by exposure to environmental mutagens for most types of damage will be permanent. Our society therefore must make decisions that in effect might determine the genetic disease burden of all human generations to come. This is an awesome challenge, particularly to the toxicologic community.

The most prudent course of action regarding human genetic risk would be to:

1. Continue to use the best available animal tests for genetic risk analysis and promote the development of more sensitive methods to estimate genomic changes in experimental animals as well as humans.
2. View chemicals that produce positive responses in one or more of the germ cell assays as probable human mutagens.
3. Make human risk calculations based on animal studies using conservative (linear, nonthreshold) models.
4. Devote greater effort to define and quantify relationships between alterations easily measured in human somatic cells (SCE, DNA adducts, repair patches, etc.) and germ cell impact.

For additional views about human genetic risk assessment see Brusick *et al.*,[61] Woodhead *et al.*,[62] and Russel *et al.*[41]

## REFERENCES


1. Newcombe, H. B.: Methods of obtaining epidemiological data for mutation risk estimation. In *Handbook of Mutagenicity Test Procedures* (B. J. Kilbey, M. Legator, W. Nichols, and C. Ramel, eds.), Elsevier, Amsterdam, pp. 461–476, 1977.
2. Carter, C. O.: The relative contribution of mutant genes and chromosome abnormalities to genetic ill-health in man. In *Progress in Genetic Toxicology,* Vol. 2 (D. Scott, B. A. Bridges, and F. H. Sobels, eds.), Elsevier/North-Holland, Amsterdam, p. 114, 1977.
3. Russell, L. B.: Validation of the *in vivo* somatic mutation method in the mouse as a prescreen for germinal point mutations. *Arch. Toxicol.* **38:**75–85, 1977.
4. Sankaranarayanan, K.: Protection against genetic hazards from environmental chemical mutagens: Experience with ionizing radiation. In *Progress in Genetic Toxicology,* Vol. 2 (D. Scott, B. A. Bridges, and F. H. Sobels, eds.), Elsevier/North-Holland, Amsterdam, pp. 77–93, 1977.

5. Hemminki, K., Sorsa, M., and Vainio H.: Genetic risks caused by occupational chemicals. *Scand. J. Work Environ. Health* **5**:307–327, 1979.
6. Sram, R. J., and Kuleshov, N. P.: Monitoring the occupational exposure to mutagens by the cytogenetic analysis of human peripheral lymphocytes *in vivo*. *Arch. Toxicol.* **1980**(Suppl. 4):11–18, 1980.
7. Bloom, A. D. (ed.): *Guidelines for Studies of Human Populations Exposed to Mutagenic and Reproductive Hazards*, March of Dimes Birth Defects Foundation, New York, 1981.
8. Committee on Mutagenicity of Chemicals in Food, Consumer Products, and the Environment. *Guidelines for Testing of Chemicals for Mutagenicity*, Department of Health and Social Security, Great Britain, 1979.
9. Matsunaga, E.: Perspectives in mutation epidemiology 1. Incidence and prevalence of genetic disease (excluding chromosomal aberrations) in human populations. *Mutat. Res.* **99**:95–128, 1982.
10. Nebert, D. W., Levitt, R. C., and Pelkonen, O.: Genetic variation in metabolism of chemical carcinogens associated with susceptibility to tumorigenesis. In *Carcinogens: Identification and Mechanisms of Action* (A. C. Griffin and C. R. Shaw, eds.) Raven Press, New York, pp. 157–185, 1979.
11. Paterson, M. C.: Environmental carcinogenesis and imperfect repair of damaged DNA in *Homo sapiens:* Causal relation revealed by rare hereditary disorders. In *Carcinogens: Identification and Mechanisms of Action* (A. C. Griffen and C. R. Shaw, eds.), Raven Press, New York, pp. 251–276, 1979.
12. Generoso, W. M., Cain, K. I., and Bandy, A. J.: Some factors affecting the mutagenic response of mouse germ cells to chemicals. In *Utilization of Mammalian Specific Locus Studies in Hazard Evaluation and Estimation of Genetic Risk* (F. J. de Serres and W. Sheridan, eds.), Plenum Press, New York, pp. 227–239, 1983.
13. Baker, T. G.: Effects of ionizing radiations on mammalian oogenesis: A model for chemical effects. *Environ. Health Perspect.* **24**:3137, 1978.
14. Sobels, F. H.: The parallelogram: An indirect approach for the assessment of genetic risks from chemical mutagens. In *Progress in Mutation Research*, Elsevier/North-Holland, Amsterdam, pp. 323–327, 1982.
15. Russell, L. B., and Montgomery, C. S.: Supermutagenicity of ethylnitrosourea in the mouse spot test. Comparisons with methylnitrosourea and ethylnitrosourethane. *Mutat. Res.* **92**:193–204, 1982.
16. Ames, B. N.: Dietary carcinogens and anticarcinogens. *Science* **221**:1256–1264, 1983.
17. Shamberger, R. J., Corlett, C. L., Beaman, K. D., and Kasten, B. L.: Antioxidants reduce the mutagenic effect of malonaldehyde and β-propiolactone. Part IX, Antioxidants and cancer. *Mutat. Res.* **66**:349–355, 1979.
18. Abrahamson, S., Bender, M. A., Conger, A. D., and Wolff, S.: Uniformity of radiation-induced mutation rates among different species. *Nature* **245**:460–462, 1973.
19. Schalet, A. P., and Sankaranarayanan, K.: Evaluation and reevaluation of genetic radiation hazards in man. I. Interspecific comparison of estimates of mutation rates. *Mutat. Res.* **35**:341–370, 1976.
20. National Academy of Sciences: *Risk Assessment in the Federal Government: Managing the Process*, National Academy Press, Washington, D.C., 1983.
21. De Serres, F. J.: Report of the Conference on Modification of Mutagenic and Carcinogenic Activity. *Mutat. Res.* **54**:197–202, 1978.
22. Sugimura, T.: Let's be scientific about the problem of mutagens in cooked food. *Mutat. Res.* **55**:149–152, 1978.
23. Vogel, E.: The relation between mutational pattern and concentration by chemical mutagens in *Drosophila*. In *Screening Tests in Chemical Carcinogenesis* (H. Bartsch, R. Montesano, and L. Tomatis, eds.), IARC Scientific Publication No. 12, pp. 117–132, 1976.

24. Dean, B. J., and Senner, K. R.: Detection of chemically-induced somatic mutation in Chinese hamsters. *Mutat. Res.* **46**:403–407, 1977.
25. Dean, B. J., and Senner, K. R.: Detection of chemically-induced mutations in tissues of Chinese hamsters. In *Progress in Genetic Toxicology,* Vol. 2 (D. Scott, B. A., Bridges, and F. H. Sobels, eds.), Elsevier/North-Holland, Amsterdam, pp. 201–206, 1977.
26. Strauss, G. H., and Albertini, R. J.: 6 Thioguanine resistant lymphocytes in human peripheral blood. In *Progress in Genetic Toxicology,* Vol. 2 (D. Scott, B. A. Bridges, and F. H. Sobels, eds.), Elsevier/North-Holland, Amsterdam, pp. 327–334, 1977.
27. Russell, L. B., and Major, M. H.: Radiation-induced presumed somatic mutations in the mouse. *Genetics* **42**:161–175, 1957.
28. Russell, L. B.: Somatic cells as indicators of germinal mutations in the mouse. *Environ. Health Perspect.* **24**:113–116, 1978.
29. Styles, J. A., and Penman, M. P.: The mouse spot test: An appraisal of its performance in detecting chemical carcinogens and mutagens. *Mutat. Res.* **154**:183–204, 1985.
30. Fabro, S.: Penetration of chemicals into the oocyte, uterine fluid, and preimplantation blastocyst. *Environ. Health Perspect.* **24**:25–29, 1978.
31. Loveless, A.: Possible relevance of O6 alkylation of deoxyguanosine to the mutagenicity and carcinogenicity of nitrosamines and nitrosamides. *Nature* **223**:206, 1969.
32. Malling, H. V., and Valcovic, L. R.: New approaches to detection of gene mutations in mammals. In *Advances in Modern Toxicology,* Vol. 4 (G. Flamm and M. Mehlman, eds.), Hemisphere Press, New York, pp. 149–171, 1978.
33. ICPEMC, Committee 4 Final Report. Estimation of genetic risks and increased incidence of genetic disease due to environmental mutagens. *Mutat. Res.* **115**:255–291 (1983).
34. Ehling, U. H.: Induction of gene mutations in germ cells of the mouse. *Arch. Toxicol.* **46**:123–138, 1980.
35. Ehling, U. H.: Risk estimations based on germ-cell mutations in mice. In *Environmental Mutagens and Carcinogens (Proceedings of the 3rd International Conference on Environmental Mutagens)* (T. Sugimura, S. Kondo, and H. Takebe, eds.), University of Tokyo Press, Tokyo, and Alan R. Liss, New York, pp. 709–719, 1982.
36. Russell, L. B., Selby, P. B., vanHalle, E., Sheridan, W., and Valcovic, L.: The mouse specific locus test with agents other than radiations: Interpretation of data and recommendations for future work. *Mutat. Res.* **86**:329–354, 1981.
37. Kratochvilova, J., and Ehling, U. H.: Dominant cataract mutations induced by γ-irradiation of male mice. *Mutat. Res.* **63**:221–223, 1979.
38. Selby, P. B., and Lee, S. S.: Sensitive-indicator results show that ethylnitrosurea is also a super mutagen for dominant skeletal mutations. *Environ. Mutagen.* **3**:373, 1981.
39. Adler, I. D.: Comparison of types of chemically induced genetic changes in mammals. *Mutat. Res.* **115**:293–321, 1983.
40. Generoso, W., Huff, S. W., and Cain, K. T.: Relative rates at which dominant-lethal mutations and heritable translocations were induced by alkylating chemicals in post-meiotic male germ cells of mice. *Genetics* **93**:163–171, 1979.
41. Russell, L. B., Aaron, C. S., de Serres, F., Generoso, W. M., Kannan, K. L., Shelby, M., Springer, J., and Voytek, P.: Evaluation of mutagenicity assays for purposes of genetic risk assessment. *Mutat. Res.* **134**:143–157, 1984.
42. Lewtas, J.: A quantitative cancer risk assessment methodology using short-term genetic bioassays: The comparative potency method. In *Risk and Reason: Risk Assessment in Relation to Environmental Mutagens and Carcinogens* (P. Oftedal and A. Brogger, eds.), Alan R. Liss, New York, pp. 107–120, 1986.
43. Cuddihy, R. G., McClellan, R. O., and Griffith, W. C.: Variability in target organ deposition among individuals exposed to toxic substances. *Toxicol. Appl. Pharmacol.* **49**:179–187, 1979.

44. Lee, I. P., and Dixon, R. L.: Factors influencing reproduction and genetic toxic effects on male gonads. *Environ. Health Perspect.* **24**:117–127, 1978.
45. Lee, W. R.: Dosimetry of chemical mutagens in eukaryote germ cells. In *Chemical Mutagens: Principles and Methods for Their Detection,* Vol. 5 (A. Hollaender and F. J. de Serres, eds.), Plenum Press, New York, pp. 177–202, 1978.
46. Russell, W. L., Huff, S. W., and Gottlieb, D. J.: The insignificant rate of induction of specific locus mutations by five alkylating agents that produce high incidences of dominant lethality. *Biol. Div. Annu. Prog. Rep. ORNL* **4535**:122–123, 1969.
47. Würgler, F. E., Sobels, F. H., and Vogel, E.: *Drosophila* as assay system for detecting genetic changes. In *Handbook of Mutagenicity Test Procedures* (B. Kilbey, C. Ramel, M. Legator, and W. Nichols, eds.), Elsevier/North-Holland, Amsterdam, pp. 335–373, 1977.
48. Mekler, P. H., Delehanty, J. T., Lohman, P. H. M., Brower, J., Putte, P. V. D., Pearson, P., Pouwels, P. H., and Ramel, C.: The use of recombinant DNA technology to study gene alteration. *Mutat. Res.* **153**:13–55, 1985.
49. Committee for a Study on Saccharin and Food Safety Policy: *Saccharin: Technical Assessment of Risks and Benefits, Report No. 1,* Assembly of Life Sciences/Institute of Medicine, National Research Council, National Academy of Sciences, Washington, D. C.; November 1978.
50. Wilson, J. G.: *Environment and Birth Defects,* Academic Press, New York, 1973.
51. Meselson, M., and Russell, K.: Comparisons of carcinogenic and mutagenic potency. In *Origins of Human Cancer, Book C, Human Risk Assessment,* Vol. 4 (H. H., Hiatt, J. D. Watson, and J. A. Winsten, eds.), Cold Spring Harbor Laboratory, Cold Spring Harbor, New York, pp. 1473–1481, 1977.
52. Clive, D., Johnson, K. O., Spector, J. F. S., Batson, A. G., and Brown, M. M. M.: Validation and characterization of the L5178Y/TK +/− mouse lymphoma mutagen assay system. *Mutat. Res.* **59**:61–108, 1979.
53. Ashby, J., and Styles, J. A.: Does carcinogenic potency correlate with mutagenic potency in the Ames assay? *Nature* **271**:452–455, 1978.
54. Mount, D. W., Little, J. W., Markham, B., Ginsburg, H., Yanish, C., and Edmonston, S.: Mechanisms of mutagenesis in bacteria. In *Environmental Mutagens and Carcinogens* (T. Sugimura, S. Kondo, and H. Takebe, eds.), Alan R. Liss, New York, pp. 105–111, 1982.
55. Brash, D. E., and Hart, R. W.: DNA damage and repair *in vivo. J. Environ. Pathol. Toxicol.* **2**:79–114, 1978.
56. Brash, D. E., and Hart, R. W.: Biopsy measurement of DNA damage and repair *in vivo:* Single-strand breaks, alkali-labile bonds, and endonuclease-sensitive sites. *Radiat. Res.* **91**:169–180, 1982.
57. Hart, R. W., Fu, P., and Chang, M. J. W.: Comparative removal of PAH adducts *in vivo.* In *Proceedings of the Sixth Annual Symposium on Polynuclear Aromatic Hydrocarbons,* Batelle's Columbus Laboratories, Columbus, Ohio, 1982.
58. Myhr, B., Turnbull, D., and DiPaolo, J. A.: Ultraviolet mutagenesis of normal and xeroderma pigmentosum variant human fibroblasts. *Mutat. Res.* **62**(2):341–353, 1979.
59. Seiler, J. P.: Apparent and real thresholds: A study on two mutagens. In *Progress in Genetic Toxicology,* Vol. 2 (D. Scott, B. A. Bridges, and F. H. Sobels, eds.), Elsevier/North-Holland, Amsterdam, pp. 233–238, 1977.
60. Ehling, U. H., Auerbeck, D., Cerutti, P. A., Friedman, J., Greim, H., Kolbye, A. C., and Mendelsohn, M. L.: Review of the evidence of thresholds in the induction of genetic effects by genotoxic chemicals. *Mutat. Res.* **123**:281–341, 1983.
61. Brusick, D., de Serres, F. J., Everson, R. B., Mendelsohn, M. L., Neel, J. V., Shelby, M. D., and Waters, M. D.: Monitoring the human population for mutations and DNA

damage. In *Guidelines for Studies of Human Populations Exposed to Mutagenic and Reproductive Hazards* (A. D. Bloom, ed.), March of Dimes Foundation, New York, pp. 111–140, 1981.
62. A. D. Woodhead, C. J. Shellabarger, V. Pond, and A. Hollaender (eds.): *Assessment of Risk from Low-Level Exposure to Radiation and Chemicals,* Plenum Press, New York, 1985.

CHAPTER 6

# Applications of Genetic Toxicology to Human and Environmental Monitoring

## INTRODUCTION

Techniques available to the genetic toxicologist offer some unique opportunities in the area of monitoring. The reasons for this are:

1. Many methods are cost-effective, yet very sensitive. Biological activity of some genotoxic agents can be detected *in vitro* at environmental concentrations lower than is detectable using analytical chemical techniques.
2. Monitoring techniques can be applied to direct human population evaluation.
3. Tests can be conducted with small sample volumes typical of those recoverable from environmental sampling devices (filters, precipitators, SAS trains).

Both human and environmental monitoring are active areas of research and much has been written concerning the advantages and limitations of using biological monitoring to set worker exposure limits or allowable environmental pollution levels.

## ENVIRONMENTAL MONITORING

Mutagens enter the environment from many sources. A substantial amount of organic chemicals enter as the result of fossil fuel combustion (petroleum, coal, wood). Other sources contributing substantial amounts are waste products and fugitive emissions from industrial processes and runoff from chemical storage and disposal sites. Finally, mutagenic substances are produced by plants and animals and are formed during the processing of foods and burning of tobacco. Table 6.1 identifies some of the probable sources of mutagenic organic compounds found in ambient air.

A feature of environmental monitoring is that most samples consist of mixtures of physical states (e.g., slurries) as well as mixtures of chemicals including inorganic and organic substances. For example, particles collected from ambient air are usually carbonaceous central cores with organic chemical bonds to them. Consequently, environmental samples often require several processing steps prior to testing. Figure 6.1 outlines the general categories of sample types encountered in environmental monitoring and recommendations for preparing the samples for bioassay techniques.

Because most environmental monitoring involves collection of large quantities of sample followed by concentration, extraction, fractionation, or other types of sample preparation, it is necessary to modify the response/unit dose. Bioassay results obtained from sources that are concentrated and/or fractionated prior to testing should be normalized for the pretesting processing and then expressed as units of biological activity per rate of emission or waste discharge. In addition, multiple samples from a collection site are necessary to establish the normal variation of the sampled work area or emission source.

Fractionation is one of the pretest processing steps. Generally, simple fractionation, such as isolating discrete phases (solid, liquid, etc.) from a complex matrix, is all that is required before evaluation. However, in specific situations discrete phases may require further fractionation into subfractions containing basic chemical classes, such as acid, basic, and neutral molecular species (Figure 6.1). Examples of such situations are:

1. If a sample is highly toxic, such that the maximum tolerated dose (MTD) for a bioassay is in the range 1–25 μg/ml, it might be useful to subfractionate the sample in order to separate a highly toxic component of the mixture from the remaining components, allowing for expression of moderate to weak mutagens that may be present.
2. Identity of specific genotoxic chemicals or chemical classes might be desired in a mixture. Fractionation followed by application of bioassay techniques can be useful in directing the investigator to the agents of interest. The term for this process is "biodirected fractionation" and it has been used to identify the mutagenic com-

TABLE 6.1
Particles in Ambient Air

| Source | Organic loading | Chemical/physical properties | Reported biological effects |
|---|---|---|---|
| Coal fly-ash from coal-fired power plant, home heating, and coke production<br>Wood-burning fly-ash from home heaating units | Low | A range of particle sizes, including those of respirable size; the carbonaceous core contains bound inorganic and organic materials; the level of bound organics is low, ranging from 0.1 to 1.0% by weight | Mutagenic activity has been demonstrated in prokaryotic microbial assays using organic solvent extracts of large quantities of stack-sampled particles; level of biological activity is highly dependent upon the temperature at which the particles were collected |
| Natural fly-ash (volcanic)<br>Commercial and home oil-fired heating units<br>Spark ignition gasoline combustion engines<br>Diesel engines | High | A range of particle sizes, including a high proportion of respirable size; the carbonaceous core contains bound inorganic and organic materials; the level of bound organics is high, ranging from 10 to 20% by weight | Mutagenic and transforming activity has been demonstrated for organic solvent extracts *in vitro* using target cells; organic extracts have also been shown to induce tumors in rodents under specific routes of administration (i.e., skin painting, intratracheal instillation, and subcutaneous injection); the levels of biological activity appear to vary according to combustion efficiency, atmospheric conditions during cooling, fuel composition, and, in case of mobile sources, the running conditions of the engine |

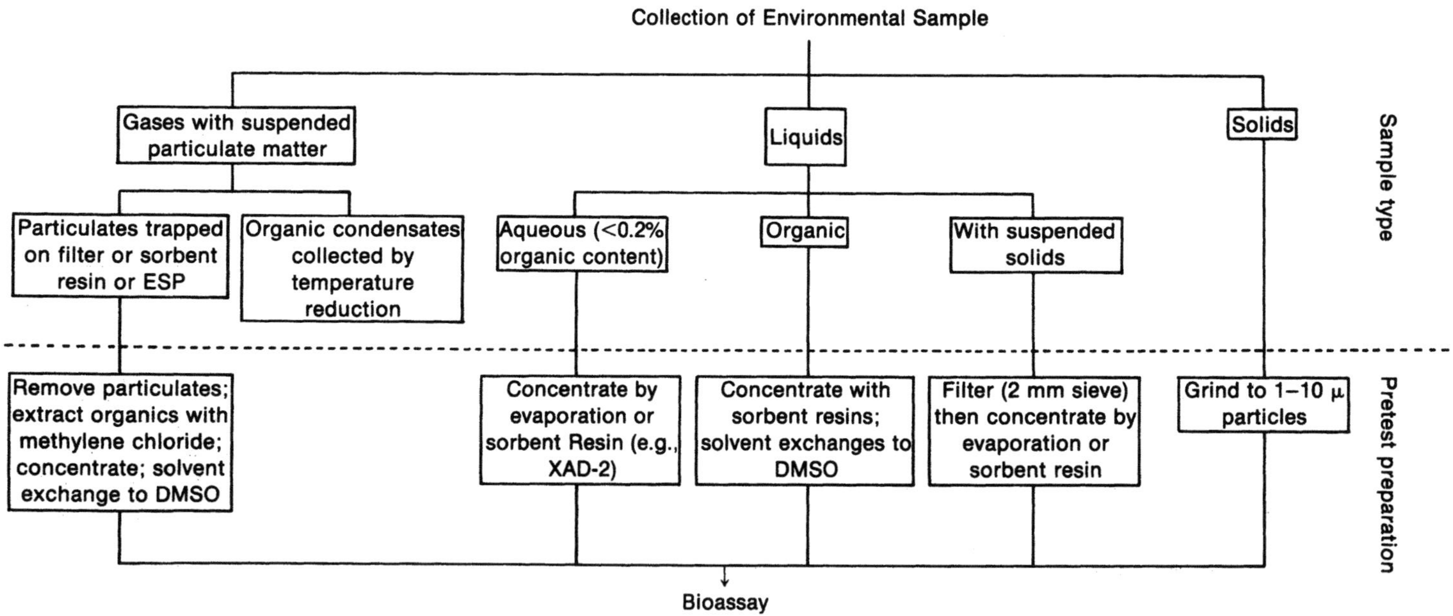


FIGURE 6.1. Summary of methods for the collecting and processing of environmental samples. Samples may also be subjected to fractionation schemes to separate acid, base, and neutral components prior to bioassay.

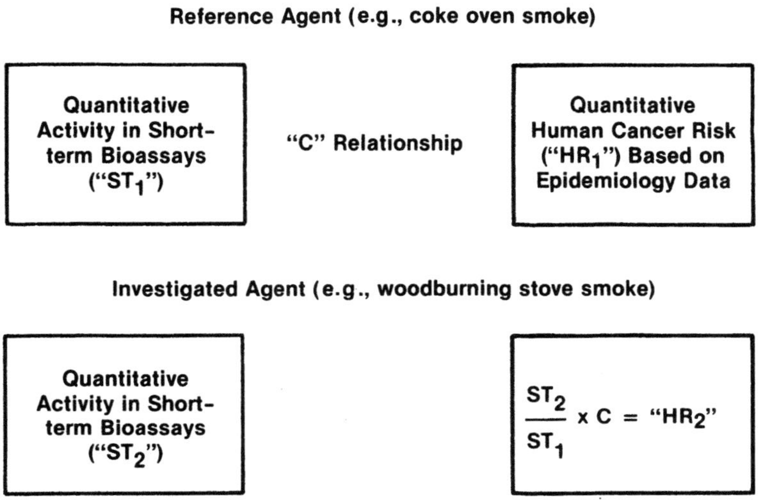


FIGURE 6.2. The use of a modified parallelogram method to estimate cancer risks from human and *in vitro* data. This method has been used by Lewtas[4] to estimate risk from environmental exposures. $C$ is a constant derived from the relative potency of the *in vitro* test for the genotoxicity of a reference agent and the human cancer risk $HR_1$ based on epidemiology. Comparisons of the same short-term test results from a similar type of environmental substance (investigated agent) can be used to calculate an estimated cancer risk $HR_2$ using the constant $C$ derived from the reference agent data.

ponents present in cooked foods,[1] photocopier toners,[2] and diesel exhaust emissions.[3]

Environmental monitoring provides an almost ideal situation for the application of short-term predictive techniques, including those for genotoxicity, for the following reasons:

1. Collection of environmental samples typically yields small volumes (after processing), which would not be suitable for larger, animal-based studies.
2. Site monitoring necessitates repeat collection and testing several times to identify variability or regularly scheduled sampling to monitor process effluents. Short-term bioassay methods are sufficiently rapid and economical to permit this type of sequence testing.
3. The reproducibility and sensitivity of some bioassays (e.g., the Ames test) are such that they can detect small variations in the toxicity of effluents brought about by changes in raw materials or processing adjustments. This level of sensitivity may be important

for the protection of highly susceptible flora and fauna in areas near discharge sites.

4. The orientation of environmental monitoring is focused toward human health effects as well as ecological toxicity and many of the test species used in short-term bioassays are relevant predictors for both concerns (i.e., bacteria, fungi, insects, plants).

Short-term bioassays have even been employed in risk assessment modeling for human health effects. Lewtas[4] has published the results of comparative risk analysis for several environmental exposures. The method consists in matching the quantitative responses from a series of short-term data with the quantitative human risk estimate derived from epidemiologic data. The risk estimate for the exposure under investigation is then made by conducting the same series of short-term tests and assuming a constant relative risk between short-term bioassays and human risk (Figure 6.2). This approach is used to estimate risks only when similarities exist between the environmental particles and the exposure conditions. For example, relative risks for roofing applicators and coke oven workers have been calculated using this approach, with good results.

## OCCUPATIONAL MONITORING APPROACHES

There are two levels of monitoring that can be followed to assess the possible effects of environmental mutagens on workers. The first approach is to monitor the immediate environment (e.g., the occupational setting) for mutagens. The other alternative is to monitor cells or body fluids of the individuals working in the environment. If mutagens are discovered in samples obtained from the workplace, controls can be applied to the source of the mutagens or personnel protective devices might be employed. In this manner, all workers will be protected to the extent that the controls or protective devices are effective. This approach is also relatively inexpensive compared to worker population monitoring.

Monitoring individual workers has been proposed to compensate for individual heterogeneity and in situations where exposure has already occurred in the absence of protective measures.

Exposure of a group of individuals to a chemical may result in a wide range of responses, depending on the lifestyle of the individuals (smokers versus nonsmokers), genetic predisposition (high-risk individuals), and work habits (conscientious use of protective equipment). Individual monitoring for exposure using noninvasive or mildly invasive sample collecting is considerably more costly than general monitoring of the work environment, but might be necessary in "after-the-fact" exposure situations.

Adjustment for nonoccupational factors is important to interpret results, since data have been published showing that smokers have higher levels of urine mutagens,[5] SCEs[5], and abnormal sperm[6] than do nonsmokers, and that the consumption of cooked beef increases the level of fecal mutagens.[7] These publications not only demonstrate the general applicability of genetic monitoring techniques, but they clearly point out the need to develop well-defined control baseline data for human populations prior to their application in monitoring subjects in the workplace.

## Occupational Monitoring Objectives

Certain occupational groups are exposed at much higher concentrations than is the general population chronically or acutely (accidents) to potentially hazardous genotoxins. The possibility exists that these exposures may significantly increase the risk of reproductive disease and/or cancer for some of the workers.[8]

Epidemiologic studies of exposed populations have long been the standard means of assessment of occupationally related disease. However, effectiveness of the epidemiologic approach to define genetic toxicity is limited because:

1. Genetic damage to the gene pool is irreversible and the nature of epidemiologic studies is retrospective rather than prospective.
2. Genetic alterations are rare population events and there is generally limited access to an adequate number of exposed individuals for reliable analysis.
3. Long latent or expression periods between exposure and onset of clinically diagnosed disease are common for genetic effects and preclude clear cause-and-effect observations.
4. Medical or death records may be inadequate.
5. Appropriate control cohorts may not exist.
6. Costs associated with epidemiologic studies are enormous.

Ideally, an occupational monitoring technique should provide sensitive, reliable, and quantitative information regarding "biologically significant" exposure to hazardous agents. "Biologically significant" assumes that the interaction or endpoint observed is a reliable indication that adverse health effects are possible and/or likely. A review of general methods has been published by de Serres.[9] Some of the newer techniques have the capacity to provide direct assessment of genes selected for their known involvement in human disease.

# CYTOGENETIC TECHNIQUES

## Chromosome Damage and Cancer

Since chromosome analysis has been the most commonly used technique, it will be of value to review some of the data that support the application of cytogenetics to the assessment of cancer in human populations. Much of the information that supports a relationship between chromosome (DNA) alterations and cancer does not come from studies of induced chromosome damage, but from inherited susceptibility to cytogenetic damage. Rare hereditary disorders in humans result in individuals who are at a significantly higher risk to cancer and also show chromosome instability and fragility as well as inherited defective DNA repair processes.[10] Examples of several of these syndromes and their effect on tumor incidence in the affected populations are shown in Table 6.2. Slightly elevated levels of SCEs have also been reported in the cancer-prone syndrome dyskeratosis congenita.[12] The concordance between chromosome structure instability and cancer appears to involve a cause–effect relationship involving very specific genetic disruption. Cells from cancer patients often contain repetitive chromosome rearrangements and an association of particular chromosome rearrangements with specific

TABLE 6.2
Association of DNA Repair Deficiencies and Chromosome Fragility in Humans with Cancer Susceptibility[a]

| | | Cancer susceptibility | |
|---|---|---|---|
| Trait | Deficiency | Type | Rate |
| Xeroderma pigmentosum (homozygous recessive) | DNA repair; probably excision type | Skin[b] | >0.5 |
| | | Melanoma[b] | 0.1 |
| Ataxia telangiectasia (homozygous recessive) | DNA repair; "X-ray" type | All types | 0.1 |
| | | Leukemia[b] | 0.02 |
| | | Lymphoreticular[b] | 0.06 |
| Fanconi's anemia (homozygous recessive) | DNA repair; cross-link type | All types | >10-fold normal |
| Bloom's syndrome (homozygous recessive) | Chromosome fragility; DNA synthesis suppression | All types | 0.25 |

[a] Adapted from Setlow[11] and Dr. J. German (unpublished results).
[b] Approximate normal cancer probabilities: skin, $5 \times 10^{-3}$/year; melanoma, $6 \times 10^{-5}$/year; lymphoreticular, $13 \times 10^{-5}$/year; leukemia, $42 \times 10^{-6}$/year.

TABLE 6.3
Nonchromosomal Genetic Monitoring Techniques

| | |
|---|---|
| Mutagens detected in body fluids: blood, urine, feces | Body fluids used as test materials in genetic assays such as the Ames test; extraction and concentration methods routinely employed |
| Lymphocyte damage: DNA strand breakage, unscheduled DNA synthesis, specific-locus mutation (HGPRT) | Techniques capable of measuring DNA damage and gene mutation in peripheral lymphocytes |
| Alkylation of hemoglobin molecules | Covalent binding to protein |
| Detection of variant hemoglobin molecules | Presumed gene mutation |
| Human abnormal spermheads | Alterations in spermhead morphology resulting from genetic damage |
| DNA adduct formation | Covalent binding to DNA |
| DNA hybridization | DNA probes to recognize altered sequences |
| Restriction fragment mapping | Changes in restriction fragment lengths due to altered DNA sequences |

types of malignancy. For example, Burkitt's lymphoma (BL) patients all show a chromosome translocation. For BL and several of these translocation/cancer associations oncogene sites have been mapped at areas of the chromosomes involved in the chromosome alterations.[13]

Cytogenetic analysis of peripheral blood lymphocytes has been used as a biological monitor for human exposure to genotoxic substances. Chromosomal aberrations constitute one class of mutation events. Their induction in blood cells indicates the possibility of risk to somatic cells and implies risk to germinal tissue, which would impact on reproductive capacity and cause genetic diseases.[14–17]

More recently, sister chromatid exchange (SCE) has been introduced as a human monitoring technique.[18] SCE analysis appears to be more sensitive than aberration analysis for detecting induced effects for some chemicals, although, in contrast to chromosome aberrations, SCEs are not known to be directly associated with mutation induction. Large numbers of mutagens and clastogens, however, have been shown to induce SCEs and there is a correspondence between SCE and point mutation induction by some chemicals *in vitro*.[19] SCE, in addition to its high degree of sensitivity, is also less subjective and simpler to score than aberrations.

Other monitoring techniques besides cytogenetic screening have been proposed (Table 6.3). While some of these tests may yield useful information, the majority have not yet reached the stage of routine application to occupational screening situations. The only exceptions to this generalization are the analysis of human urine concentrates for the presence of mutagenic substances and the analysis for DNA adducts.

This urine analysis technique has proven useful in several monitoring studies by detecting bioactivated molecules excreted directly or as conjugates in human urine.[20,21]

In summary, the present state of the art encompasses three techniques ready for application to the problem of detecting human exposures to genotoxic substances. Chromosome analysis is the monitoring technique with the greatest use experience. SCE analysis and urine testing are both relatively new techniques, but offer obvious advantages in sensitivity and performance time.

## INTERPRETATION OF HUMAN MONITORING RESULTS

Cytogenetic studies of populations have proven to be extremely useful in biological dosimetry for individuals exposed to genotoxic agents. Increased incidences of chromosome aberrations and/or SCEs are indicators of "biologically significant" exposure, whereas negative findings do not necessarily rule out some exposure.

The critical question that must be addressed in the interpretation of human monitoring studies is the meaning of positive findings. One can show that somatic cells are at risk for chromosome mutation, but can this observation be extrapolated to other types of toxicologic endpoints? Are increases in chromosome aberrations biological predictors for other mutagenic and carcinogenic effects? Definite proof for establishing the link between chromosomal changes in exposed populations and cancer, genetic disease, or reproductive defects will require carefully conducted epidemiologic analyses. Table 6.4 gives a list of human studies that support an association between increases in chromosome alterations and the eventual production of cancer following exposure to certain agents in normal use or occupational settings.

Analysis of urine for mutagens may be conducted as a supplementary test for human exposure. The interpretation of results obtained by this technique, however, must be approached with caution. Presence of mutagens in the urine can be considered evidence of exposure to a mutagenic chemical or to a chemical that forms mutagenic metabolites that are eventually excreted. However, excretion of mutagens may be a protective process, especially if the metabolites are conjugated and thus are in a biologically inactive form. In fact, the absence of mutagens in the urine could be interpreted as evidence that active agents encountered in low concentrations are completely covalently bound to cellular macromolecules (i.e., potentially causing damage) and thus are not available for

TABLE 6.4
Association between Induced Chromosome Alterations and Cancer in Humans[a]

| Agent or occupation associated with chromosome damage | Known human effect | Chromosomal endpoint detected at occupation or use exposure conditions | Reference |
|---|---|---|---|
| Radiation | Carcinogen under occupational exposure | Chromosome aberrations; statistically significant increases | 22 |
| Cigarette smoke | Carcinogen under use condition | Sister chromatid exchange (SCE); statistically significant increases | 23 |
| Vinyl chloride | Carcinogen under occupational exposure | Chromosome and SCE increases statistically significant | 24 |
| Cyclophosphamide | Carcinogen under use conditions | Chromosome aberrations; statistically significant increases | 25 |
| Melphan | Carcinogen under use conditions | Statistically significant increases in SCE under use conditions | 26 |
| Tire and rubber industry | Carcinogen under occupational exposure | Increase in chromosome aberrations in lymphocytes of workers; 2.63% compared to 1.14% in controls | 27 |
| Benzene | Probable leukemogen under occupational exposure | Increased chromosome aberrations in motor fuel handlers; statistical analysis not described; other studies show no increases; new IARC monograph will report positive for alterations | 28 |
| Arsenic | Carcinogen under occupational exposure | Increases chromosome aberrations; good correlation with exposure levels in workers at a smelter in Sweden | 29 |
| Chromium compounds | Carcinogen under occupational exposure | Increased chromosomal aberrations seem to be statistically significant | 30 |
| Benzidine | Carcinogen under occupational exposure | Statistically significant increase in SCE in workers exposed | 31 |
| *Bis*-chloromethyl-ether | Carcinogen under occupational exposure | Chromosome aberrations increased in a sample of 12 workers exposed | 32 |

*(continued)*

TABLE 6.4 (*Continued*)

| Agent or occupation associated with chromosome damage | Known human effect | Chromosomal endpoint detected at occupation or use exposure conditions | Reference |
|---|---|---|---|
| Styrene | Roden carcinogen and suspect human carcinogen | Increases in chromosome aberrations found among workers in styrene manufacturing | 33 |
| Epichlorohydrin | Rodent carcinogen and suspect human carcinogen | Increased chromosome aberrations in occupationally exposed | 34 |
| Ethylene oxide | Rodent carcinogen and suspect human carcinogen | Increased chromosome aberrations and SCEs in workers | 35–37 |
| Cadmium | Carcinogen under occupational exposure | Nonsignificant increases in chromosome aberrations observed among workers exposed to cadmium | 38 |
| Nickel | Suspect human carcinogen under occupational exposure | Nonsignificant increases in chromosome aberrations observed among workers exposed to nickel in refinery | 39 |
| Toluene | Liver toxicity | Significant increases in SCEs and chromosome aberrations in 20 male workers exposed for more than 16 years at a rotogravure plant | 40 |

[a] The data cited above have not been subjected to a thorough review for reliability and in all cases the conclusion of the author was accepted.

urinary excretion. Consequently, knowledge of the pharmacology and metabolic fate of the suspected mutagens is essential to the proper interpretation of this monitoring technique. An additional problem is that increases known to result from cigarette smoking might mask weak responses from other etiologic agents.

While these comparisons are encouraging and suggest that population monitoring for cytogenetic alterations may be an early signal for carcinogenesis, many problems remain:

1. It is not usual to obtain clear dose relationships in human populations (often because of the lack of adequate exposure information).

2. Application of statistical analysis to data assessment is not well developed for human cytogenetic studies.
3. Selection of exposed and control groups has not been standardized, nor has a standard questionnaire been designed to obtain relevant personal histories.
4. The length of time that cytogenetic effects persist following exposure is not well established, particularly in the case of SCE.

Up to this point, the discussion has centered on risk to somatic cells. Of equal or greater concern for exposed individuals is the integrity of genes in germ cells. One cannot directly extrapolate clastogenic effects from somatic cell to germ cell, because of inadequate information. Human population studies in which chromosome damage was clearly established in somatic cells have not been accompanied by the appropriate epidemiologic studies to demonstrate heritable genetic effects. Studies with animals suggest that germ cell responses are significantly less than somatic cell responses to mutagenic agents.

In spite of the recognized limitations of this technology, early efforts clearly suggest the value of cytogenetic screening as a human monitor and possible early signal of damage leading to irreversible toxic effects. Human monitoring studies may also be useful in detecting subsets within the population with a high susceptibility to induced disease. As described above, several hereditary diseases that affect DNA repair and/or chromosome stability are associated with extraordinarily high cancer risk. Identification of individuals within the screened population who have extremely high levels of induced aberrations or SCEs may prove to be a valuable diagnostic tool for detecting those subsets.

## APPLICATION OF BIOTECHNOLOGY TECHNIQUES TO HUMAN MONITORING

Although the conventional techniques have proved to be valuable monitoring tools, they are sometimes limited from wider application because of the time and costs associated with their use. It is also difficult to make quantitative human hazard predictions from the results of conventional methods other than in the most general terms.

Since 1980 there has been rapid development and application of molecular analysis methods. Use of these techniques is particularly relevant to human monitoring for genetic damage and has resulted in the subdiscipline of molecular epidemiology. The technology attempts to measure directly the biologically effective dose of a genotoxic agent in exposed individuals rather than extrapolate from ambient concentration levels.

TABLE 6.5
Benzo[*a*]pyrene–DNA Adducts in Humans by Immunoassay (Polyclonal)[a]

| Populations studied | Tissue | Number positive |
|---|---|---|
| Roofers | Lymphocytes | 7/28 |
| Foundry workers | Lymphocytes | 7/20 |
| Smokers/nonsmokers | Lung tissue | 7[b]/23 |
| Volunteers | Peripheral alveolar macrophages | 3[b]/5 |
| Lung cancer patients | Lung tissue | 4/14 |
| Lung cancer controls | Lung tissue | 0/13 |

[a] From Perera *et al.*[41]
[b] All smokers.

A second aspect of molecular dosimetry is that of extrapolation. Once a molecular dosimeter (e.g., DNA adducts) has been applied to exposed humans once, it can compare those binding profiles and efficiencies with data from *in vitro* cell culture methods and/or animal models to provide data that may improve extrapolation techniques. Most of the molecular epidemiologic biomonitoring techniques available have not been thoroughly evaluated in human trials. Results from pilot and preliminary studies, however, appear very encouraging.

## Immunoassays

Use of immunoassay employing poly- or monoclonal antibodies to recognize DNA adducts appears to be one of the most promising molecular monitoring techniques. Immunoassays are extremely sensitive (the level of detectability is one adduct per $10^7$ nucleotides) and can be conducted on easily accessible human cell explants, such as lymphocytes.[41] Procedures to develop the necessary reagents for immunoassays are rather extensive and time-consuming since individual DNA-bound adducts must be isolated, characterized, and used to raise antibodies. Ensuring that all possible DNA adducts are recovered to prepare the reagents requires exhaustive DNA analysis and sophisticated analytical chemistry capability. Once antibodies with good specificity are prepared, analysis of human material is generally straightforward. Antibodies may be labeled using fluorescent or radioactive tracers.[42]

Table 6.5 illustrates the results of some preliminary studies looking for the presence of benzo[*a*]pyrene adducts in human tissues using polyclonal antibodies. Lohman and co-workers have shown detection of *N*-

acetoxy-2-acetylaminofluorene adducts at the single-cell level using digitized, computer-enhanced imaging.[43]

### $^{32}$P-Postlabeling

Immunoassays as described above are dependent upon characterization of specific DNA adducts. Another technique, the $^{32}$P-postlabeling assay, can be used to screen for adduct formation where the configuration of the adducts is unknown.[44] This is a very sensitive method, with a sensitivity to detect one altered base in $10^{10}$ unaltered nucleotides. After exposure, 1–5 μg of DNA is extracted, purified, and digested to deoxyribonucleoside-3′-monophosphates. The monophosphates are then post-treatment-labeled via incubation with γ-$^{32}$P-labeled ATP and phage T4 polynucleotide kinase. This attaches $^{32}$P to the monophosphates. The mixture of $^{32}$P-labeled unaltered nucleotides and nucleotide adducts is subjected to thin-layer chromatography (TLC), followed by autoradiography. The result is a "fingerprint" of DNA bases and adducts seen as spots located at various points on the plate. Both the amount and total number of DNA adducts can be estimated with this method. In fact, this method may be a first step in isolating specific adducts for use in preparing antibodies in the immunoassays.

### DNA Probes

Probably one of the most powerful biomonitoring techniques to emerge is the use of synthetic nucleotide polymers capable of recognizing altered gene nucleotide sequences. Specific genes can be isolated and completely sequenced. From the gene sequence a set of probes is made to rapidly screen DNA isolated from exposed individuals for alterations in or loss of bases. At the present time this technique has not been directly applied to population monitoring.

## SUMMARY

In summary, the results developed from human population monitoring studies permit only a few relatively narrow conclusions:

1. Detection of significant levels of chromosome aberration or SCE increases in peripheral blood lymphocytes is a useful dosimeter of somatic cell exposure to DNA-altering agents.
2. Somatic cell genetic effects cannot be presumed to reflect germ cell effects, but do imply a risk.

3. Populations exposed to potentially hazardous chemicals and which show significant increases in chromosome aberrations appear to be at higher risk for genetic disease, cancer, or reproductive defects than are unexposed populations.
4. Any presumption of increased risk developed on the basis of population monitoring applies to the group sampled as a whole and there are presently no methods to identify individuals within the exposed population whose health is more likely than others' to be affected.
5. Nonexposure or safety cannot be proved conclusively on the basis of negative results from a human monitoring study. This is true regardless of the genetic endpoint being monitored.

Genetic monitoring appears to be a strong candidate for early detection of populations that may be at higher risk to genetic or neoplastic diseases.

## REFERENCES


1. Hargraves, W. A., and Pariza, M. W.: Purification and mass spectral characterization of bacterial mutagens from commercial beef extract. *Cancer Res.* **43**:1467–1472, 1983.
2. Rosenkranz, H. S., McCoy, E. C., Saunders, D. R., Butler, M., Kiriazides, D. K., and Mermelstein, R.: Nitropyrenes: Isolation, identification and reduction of mutagenic impurities in carbon black and toners. *Science* **209**:1039–1043, 1980.
3. Huisingh, J. L., Bradow, R. L., Jungers, R. H., Claxton, L., Zweidinger, R. B., Tejada, S. B., Bumgarner, J., Duffield, F., Waters, M. D., Simmon, V. F., Hare, C., Rodriguez, C., and Snow, L.: Application of bioassay to the characterization of diesel particle emission. In *Application of Short-Term Bioassays in the Fractionation and Analysis of Complex Environmental Mixtures* (M. Waters, J. Hersingh, S. Sandhu, and L. Claxton, eds.), EPA Health Effects Research Laboratory, Research Triangle Park, North Carolina, pp. 1–32, 1978.
4. Lewtas, J.: A quantitative cancer risk assessment methodology using short-term genetic bioassays: The comparative potency method. In *Risk and Reason: Risk Assessment in Relation to Environmental Mutagens and Carcinogens* (P. Oftedal and A. Brogger, eds.), Alan R. Liss, New York, 1986, pp. 107–120.
5. Bridges, B. A., Clemmesen, J., and Sugimura, T.: Cigarette smoking—Does it carry a genetic risk? International Commission for Protection Against Environmental Mutagens and Carcinogens. Publication No. 3, *Mutat. Res.* **64**:71–81, 1979.
6. Kilian, D. J., and Picciano, D.: Cytogenetic surveillance of industrial populations. In *Chemical Mutagens: Principles and Methods for Their Detection*, Vol. 4 (A. Hollaender, ed.), Plenum Press, New York, pp. 321–339, 1976.
7. Bruce, W. R., Varghese, A. J., Furrer, R., and Land, P. C.: A mutagen in the feces of normal humans. In *Origins of Human Cancer*, Vol. 4 (H. H. Hiatt, J. D. Watson, and J. A. Winsten, eds.), Cold Spring Harbor Conferences on Cell Proliferation, Cold Spring Harbor Laboratory, Cold Spring Harbor, New York, pp. 1641–1646, 1977.
8. Brusick, D., de Serres, J., Everson, R., Mendelsohn, M., Neel, J., Shelby, M., and Waters, M.: Monitoring the human population for mutagenic effects: Detection of gene

mutations and DNA damage. In *Guidelines for Studies of Human Populations Exposed to Mutagenic and Reproductive Hazards* (A. D. Bloom, ed.), March of Dimes Birth Defects Foundation, pp. 111–140, 1981.
9. De Serres, F. J.: Population monitoring: Methods and applications. *Mutat. Res.* **130**:73–77, 1984.
10. LeBeau, M. M.: Chromosomal fragile sites and cancer-specific rearrangements. *Blood* **67**:849–858, 1986.
11. Setlow, R. B.: Repair deficient human disorders and cancer. *Nature* **271**:713–717, 1978.
12. Burgdorf, W., Kurvink, K., and Cervenka, J.: Sister chromatid exchange in dyskeratosis congenita lymphocytes, *J. Med. Genet.* **14**:256–257, 1977.
13. Rowley, J. D.: Human oncogene locations and chromosome aberrations. *Nature* **27**:290–291, 1983.
14. Hemminki, K., Sorsa, M., and Vainio, H.: Genetic risks caused by occupational chemicals, *Scand. J. Work Environ. Health* **5**:307–327, 1979.
15. Natarajan, A. T., and Obe, G.: Screening of human populations for mutations induced by environmental pollutants: Use of human lymphocyte system. *Ecotoxicol. Environ. Safety* **4**:468–481, 1980.
16. Sorsa, M.: Cytogenetic methods in the detection of chemical carcinogens. *J. Toxicol. Environ. Health* **6**:1077–1080, 1980.
17. Sram, R. J., and Kuleshov, N. P.: Monitoring the occupational exposure to mutagens by the cytogenetic analysis of human peripheral lymphocytes *in vivo*. *Arch. Toxicol* (Suppl. 4)**1980**:11–18, 1980.
18. Stolley, P. D., Soper, K. A., Galloway, S. M., Nichols, W. W., Norman, S. A., and Wolman, S.: Sister chromatid exchanges in association with occupational exposure to ethylene oxide. *Mutat. Res.* **129**:89–102, 1984.
19. Carrano, A. V., Thompson, L. H., Lindl, P., and Minkler, A.: Sister chromatid exchange as an indicator of mutagenesis, *Nature* **271**:551–553, 1978.
20. Falck, K., Sorsa, M., Vanio, H., and Kilpikari, V.: Mutagenicity in urine of workers in rubber industry. *Mutat. Res.* **79**:45–52, 1980.
21. Yamasaki, E., and Ames, B. N.: Concentration of mutagens from urine by absorption with the nonpolar resin XAD-2: Cigarette smokers have mutagenic urine. *Proc. Natl. Acad. Sci. USA* **74**:3555–3559, 1977.
22. Evans, H. J., Buckton, K. E., Hamilton, G. E., and Carothers, A.: Radiation induced chromosome aberrations in nuclear-dockyard workers. *Nature* **277**:531–534, 1979.
23. Lambert, B., Lindblad, A., Nordenskjold, M., and Werelius, B.: Increased frequency of sister chromatid exchanges in cigarette smokers. *Hereditas* **88**:147–149, 1978.
24. Anderson, D., Richardson, C. R., Wight, T. M., Purchase, I. G. H., and Adams, W. G. F.: Chromosomal analyses in vinyl chloride exposed workers. Results from analysis 18 and 42 months after an initial sampling. *Mutat. Res.* **79**:151–162, 1980.
25. Kucerovä, M., Polivkovä, Z., Renéltovä, I., Kocandrle, V., and Smetanovä, J.: Dynamics of chromosomal aberrations and SCE in patients treated with Imuran and cyclophosphamide. *Mutat. Res.* **74**:173, 1980.
26. Lamberg, B., Ringborg, U., Lindblad, A., and Sten, M.: Sister-chromatid exchange in peripheral lymphocytes of human subjects: The effects of cigarette smoking and anticancer chemotherapy. *Mutat. Res.* **80**:235, 1981.
27. Aleksandrov, S. E.: Incidence of chromosome aberrations in workers of tire and rubber industries. *Genetika* **18**(1):161–163, 1982.
28. Fredga, K., Reitalu, J., and Berlin, M.: Chromosome studies in workers exposed to benzene. In *Genetic Damage in Man Caused by Environmental Agents,* Proceedings of a Conference held in Oslo, Norway, May 11–13, 1977 (K. Berg, ed.), Academic Press, New York, pp. 187–203, 1979.

29. Beckman, G., Beckman, L., Nordenson, I., and Nordenson, S.: Chromosomal aberrations in workers exposed to arsenic. In *Genetic Damage in Man Caused by Environmental Agents,* Proceedings of a Conference held in Oslo, Norway, May 11–13, 1977 (K. Berg, ed.), Academic Press, New York, pp. 205–211, 1979.
30. Levis, A. G., and Bianchi, V.: Mutagenic and cytogenetic effects of chromium compounds. In *Biological and Environmental Aspects of Chromium* (S. Langard, ed.), Elsevier/North-Holland Biomedical Press, Amsterdam, pp. 171–208, 1981.
31. Bassendowska-Karska, E.: Investigation of the mutagenic properties of certain chemical substances during occupational exposure. *Pol. Tyg. Lek.* **35**(2):53–54, 1980.
32. Zudová, Z., and Landa, K.: Genetic risk of occupational exposures to haloethers. *Mutat. Res.* **46**:242–243, 1977.
33. Norppa, H., Vaino, H., and Sorsa, M.: Chromosome aberrations in lymphocytes of workers exposed to styrene. *Am. J. Ind. Med.* **2**(3):299–304, 1981.
34. Kucerovä, M., Zhurkor, V. S., Polivkovä, Z., and Ivanova, J. E.: Mutagenic effect of epichlorohydrin, II. Analysis of chromosomal aberrations in lymphocytes of persons occupationally exposed to epichlorohydrin. *Mutat. Res.* **48**:355–369, 1977.
35. Garry, V. F., Hozier, J., Jacobs, D., Wade, R. L., and Gray, D.: Ethylene oxide: Evidence of human chromosomal effects. *Environ. Mutagen.* **1**:375–382, 1979.
36. Thiess, A. M., Schwegler, H., Fleig, F., and Stocker, W. G.: Mutagenicity study of workers exposed to alkylene oxides (ethylene oxide/propylene oxide) and derivatives. *J. Occup. Med.* **23**:343–347, 1981.
37. Yager, J. W., Hines, C. J., and Spear, R. C.: Exposure to ethylene oxide at work increases sister chromatid exchanges in human peripheral lymphocytes. *Science* **219**:1221–1223, 1983.
38. O'Riordan, M. L., Hughes, E. G., and Evans, H. J.: Chromosome studies on blood lymphocytes of men occupationally exposed to cadmium. *Mutat. Res.* **58**:305–311, 1978.
39. Waksvik, H., and Boysen, M.: Cytogenetic analyses of lymphocytes from workers in a nickel refinery. Dept. Genetics, Norwegian Radium Hosp., Oslo 3, Norway. *Mutat. Res.* **103**(2):185–190, 1982.
40. Bauchinger, M., Schmid, E., Dresp, J., Kolin-Gerreshium, J., Hauf, R., and Suhr, E.: Chromosome changes in lymphocytes after occupational exposure to toluene. *Mutat. Res.* **102**:439–445, 1982.
41. Perera, R. F., Santella, R. M., and Poirier, M. C.: Potential methods to monitor human populations exposed to carcinogens: Carcinogen–DNA binding as an example. In *Risk Quantitation and Regulatory Policy,* Banbury Report 19 (D. G. Hoel, R. A. Merrill, and F. P. Perera, eds.), Cold Spring Harbor Laboratory, Cold Spring Harbor, New York, 1985, pp. 211–229.
42. Baan, R. A., Lohman, P. H., Fichtinger-Schepman, A. M., Muysken-Schoen, M. A., and Ploem, J. S.: Immunochemical approach to detection and quantitation of DNA adducts resulting from exposure to genotoxic agents. *Prog. Clin. Biol. Res.* **207**:135–146, 1986.
43. Baan, R. A., Lansbergen, M. J., deBruin, P. A. F., Willems, M. I., and Lohman, P. H. M.: The organ-specific induction of DNA adducts in 2-acetylaminofluorene-treated rats, studied by means of a sensitive immunochemical method. *Mutat. Res.* **150**:23–32, 1985.
44. Gupta, R. C., Reddy, M. V., and Randerath, K.: $^{32}P$-labeling analysis of non-radioactive aromatic carcinogen–DNA adducts. *Carcinogenesis* **3**:1081–1092, 1982.

CHAPTER 7

# The Genetic Toxicology Laboratory

## INTRODUCTION

It is difficult to formulate specifications for genetic toxicology laboratories. Laboratories vary in scope and purpose. However, certain aspects should be rather universal in nature. Excluding considerations of personnel, probably the single most critical parameter is the one of safety. The seriousness of the type of toxicity resulting from exposure to highly mutagenic and carcinogenic agents must be factored into the types of facilities and equipment used to perform genetic studies.[1]

Because of the varied nature of the assays conducted for genetic evaluation, several types of laboratory facilities are required to perform the entire range of tests. These facilities consist of separate areas for microbiology, tissue culture, cytogenetics, biochemistry, and possibly the handling of *Drosophila*. Animal facilities will also be required if *in vivo* studies are to be conducted or microsomal activation preparations are going to be made. Support facilities for media preparation, glassware washing, and chemical storage are generally located in close proximity to the testing facility.

All laboratories should be constructed and equipped to facilitate testing under safe conditions. This means that testing areas should have limited access, and the potential hazards associated with the work should be clearly identified (Figure 7.1). Most dosing and manipulation involving chemicals should be performed under certified Class II, Type A or B hoods vented through high-efficiency particulate air (HEPA) and charcoal filters to the outside. If the dosing involves animals, a hood system of the type shown in Figure 7.2 must be located near the animal holding area.

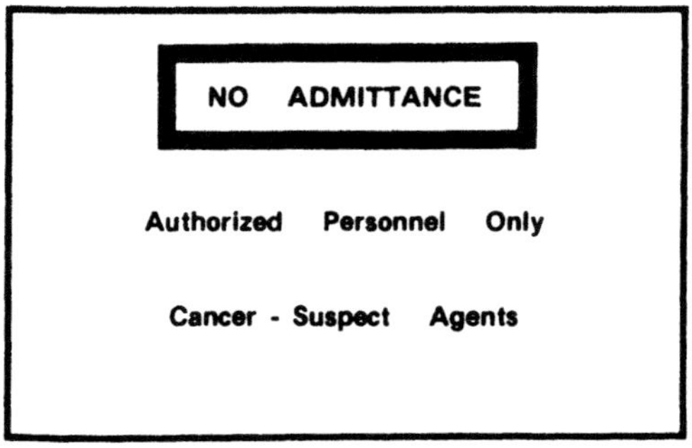


FIGURE 7.1. Warning of potential hazards. Signs such as this should be used to restrict entry into laboratory areas. Adaptations of such signs may also designate the type of safety equipment required for entry (e.g., chemical cartridge respiratory required). Use of such signs is necessary to protect technicians and nontechnical staff. The level of restrictions must be clearly stated.

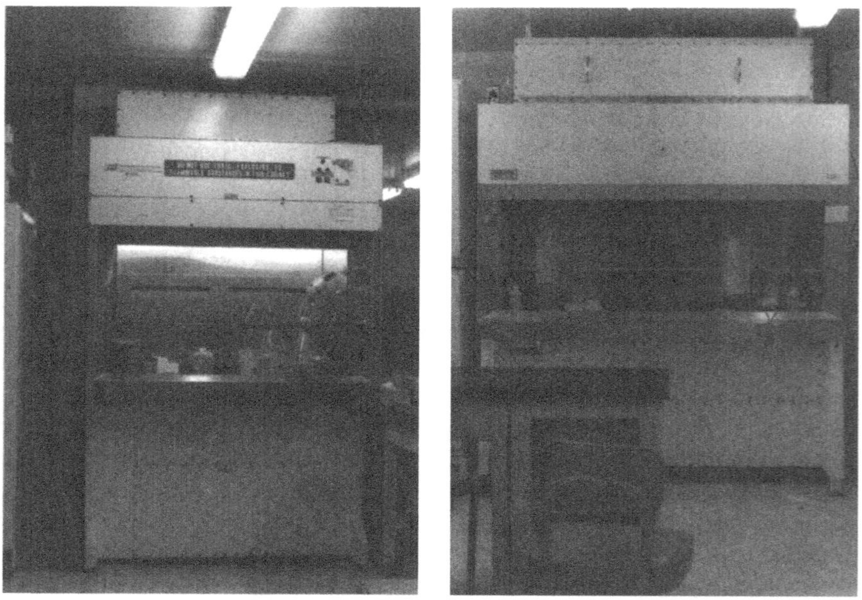

FIGURE 7.2 Examples of safety hoods for *in vitro* and animal exposures. The photograph on the left shows a Class II, Type B hood used in cell culture *in vitro* tests. The air is not recycled, but is vented through HEPA and charcoal filters. The photograph on the right shows the administration of a potentially toxic material to a small animal. This hood is also vented to protect the technician and the work area.

Employee safety equipment for controlling chemical spills or showers to rinse the eyes and skin should be readily available.

## REGULATORY REQUIREMENTS AFFECTING LABORATORY OPERATIONS

Genetic toxicology laboratory operations include numerous activities that result in human exposure to toxic, mutagenic, or carcinogenic chemicals. Even though the quantities of most of these chemicals are small, some of the compounds may be subsequently shown to be potent toxicants in animal models and may pose a hazard to humans, and regulations regarding their handling have been promulgated. Disposal of wastes generated from operations carried out in these laboratories must also comply with strict regulations. While the regulations apply to all laboratories, they are primarily focused on commercial testing laboratories or company in-house testing facilities.

### Hazardous Waste Disposal

All laboratory operations with chemicals result in the production of chemical wastes. These wastes may pose potential harm, both short and long term, to human health or to the environment unless the disposal is properly handled.

The proper handling of hazardous wastes falls under the purview of the Environmental Protection Agency (EPA). Following congressional enactment of the Resource Conservation and Recovery Act (RCRA) in 1976,[2] EPA developed a broad base of regulations, promulgated in May 1980 as 40 CFR 122-265, designed to manage hazardous wastes "from cradle to grave." These regulations impose rigid legal and moral obligations on those who generate, transport, store, treat, and dispose of hazardous wastes.

RCRA defines hazardous wastes as those demonstrating specific hazard characteristics (ignitability, corrosivity, toxicity, and reactivity) and sets limits on the quantity of those materials that may be discarded without regard to compliance. Radioactive waste disposal is not considered under RCRA, since it is regulated by the Nuclear Regulatory Commission. Biological or infectious waste generated by life science laboratories or hospitals is similarly not covered by RCRA regulations, unless it happens to be chemically contaminated also. Other federal waste effluent regulations that may impact laboratory operations are the Clean Water Act, the Clean Air Act, the Toxic Substances Control Act, and the Federal Insecticide, Fungicide and Rodenticide Act (FIFRA).

### Guidelines

In addition to the foregoing regulations, which impact the operation of the laboratory, other agencies within the public sector may require the laboratory to comply with certain rules that do not have the same force of law as a regulation, but could be considered as contractual obligations associated with performance of the grants or contracts supported by the agency. Usually these rules (or guidelines) are tailored for operations that are not specifically covered under federal laws, which are purposely written for broad, general applicability.

Three federal agencies under the Department of Health and Human Services have developed safety and health guidelines with which life science laboratories (including genetic toxicology testing laboratories) must comply. The first, the Division of Safety at the National Institutes of Health, has published guidelines providing recommendations for practices, laboratory facilities, and safety equipment in the handling of oncogenic viral agents,[3] carcinogenic chemicals,[4] and recombinant DNA.[5] Second, the Center for Disease Control jointly with the NIH has published guidelines for the control of bacterial, viral, fungal, and rickettsial infectious agents.[6] And third, the National Toxicology Program of the National Institute of Environmental Health Sciences has published guidelines for the safe conduct of animal[7] and genetic toxicology studies.[8]

## LABORATORY DESIGN AND CHEMICAL CONTAINMENT

### Work Area Containment Methodology

A wide variety of chemicals may be handled by research and testing laboratories, whether they are animal toxicology or cellular and genetic toxicology facilities. Because of the potential hazards associated with many of these chemicals, it is essential that laboratory personnel, the public, and the environment be protected by confining these materials through the use of carefully defined operating practices, specially designed facilities, and contamination control systems.

Containment of a hazard is customarily achieved through the use of well-established physical and administrative principles of contamination control. The physical controls consist of primary, secondary, and tertiary levels of containment.

1. *Primary containment,* the first line of defense for the protection of laboratory personnel and the immediate laboratory environment from exposure to hazardous materials, is provided by safety equipment, such as chemical fume hoods and biological safety

cabinets. The fundamental principle here is that the chemical is controlled at the source. The exhaust air from these devices is filtered or treated prior to discharge in order to remove the contaminant.

2. *Secondary containment,* the protection of areas external to the laboratory, is provided by the laboratory structure itself, corridor and room construction and arrangement, airlocks, a ventilation system that provides unidirectional airflows, clothing changing rooms, showers, and effluent air and water treatment systems to remove contaminants from these waste streams prior to discharge from the facility.
3. *Tertiary containment* is provided by isolating the entire barrier within a segregated or remote area of an existing facility or within a building that is physically separated from other structures.

Administrative controls relate to personnel practices, the use of personal protective clothing or equipment, handling, storage, and disposal of hazardous materials, and emergency procedures. Most institutions, including university research laboratories, have developed standard operating procedures (SOPs) describing the proper steps to protect employees and visitors in laboratory areas.

The physical and administrative controls are designed to complement each other in order to minimize the risk of exposure to workers and chemical release into the environment.

### General Laboratory Ventilation

General laboratory ventilation is designed primarily for the comfort of the employees or researchers and should not be relied upon solely to contain and control toxic gases, vapors, or dusts if they should be accidently released into a room in significant quantities. It is recommended that ventilation systems be set up to exhaust air from the laboratory to the outside and not recycle it. Work with unknown or potentially hazardous materials should be conducted in proper hood systems to avoid contact with the person engaged in the testing process.

### Laboratory Exhaust Ventilation

Laboratory hoods act to prevent exposure to toxic or flammable vapors or gases. However, the hood also serves, with the sash in a lowered position, to protect the worker from splashes, explosions, or fires which may occur, especially with reactive chemical systems, and second, as containment for accidental chemical spills. Hoods should not be used as storage areas for large quantities of chemicals or large containers, since such congestion could reduce the air flow or affect air flow patterns and thus interfere with proper hood operation.

There are several types of exhaust ventilation hoods in use in the genetic toxicology laboratory. The effectiveness of the fume hood or Class I biological cabinet relies on drawing 100% of the air through the hood and discharging it to some remote point outside the building. Since unprocessed room air is drawn into this hood, it is not normally recommended for operations where a sterile (contaminant-free) environment is required.

Exhaust air from chemical fume hoods should never be recirculated, even if the exhaust effluent is filtered or treated in some way. The reason is that unless there is a way to monitor continuously the effectiveness of the exhaust air filtration or treatment system, toxic agents may reenter the building undetected. For the same reason, exhaust air discharge should be located as far away as possible from the building's supply air intakes.

Horizontal laminar flow cabinets are frequently used in biomedical laboratories. These cabinets should *never* be used for work involving hazardous materials, since the air flows *toward* the operator, thus exposing him or her to the hazard. These cabinets should only be used for work with sterile (nonhazardous) materials.

The Class II (vertical) laminar flow biological safety cabinet is more complicated than the chemical hood and is designed to provide a microbiologically clean environment within the cabinet as well as worker protection against particulate aerosols. Class II cabinets have been certified for use by the National Sanitation Foundation under their Standard 49,[9] which sets manufacturing and performance criteria for these devices. The Class II, Type A cabinet recirculates 70% of the air through high-efficiency particulate air (HEPA) filters, across the working surface, and discharges 30% of the air (also through a HEPA filter) out of the cabinet (Figure 7.2). Type A hoods discharge the exhaust air to the laboratory room unless they have been provided with ducting to the outside. HEPA filters designed to remove 99.97% of the particulates (dusts and mists) to 0.3 nm size are used to control airborne contaminants and to remove them from exhaust airstream. HEPA filters are ineffective against gases and vapors.

There are three types of Class II, Type B cabinets. The first, a Type $B_1$ cabinet, recirculates 30% of the air and exhausts 70%. Type B hoods must be ducted to the outside of the facility and be provided with activated carbon filters on the discharged exhaust where volatile chemicals are being utilized.

Because Class II cabinets recirculate some air, large quantities of extremely volatile or highly toxic chemicals should not be used, since continued generation of the vapors may result in a gradual accumulation of material, which, in the case of a flammable liquid, could reach the flammable concentration, leading to an explosion and fire if an ignition source is present (e.g., bunsen burner, electrical arc). The second type, the Class II, Type $B_2$ biological safety cabinet, is a 100% exhaust device, which also provides a microbiologically clean working surface. This cabinet is strongly recommended for operations using substantial quantities of hazardous or unknown chemicals. The third, a Class II, Type $B_3$, is a convertible cabinet, which can serve in either Type $B_1$ or $B_2$ configurations.

The Class III hood or glove box provides the highest level of worker and environmental protection. It is totally enclosed and equipped with glove ports and is used for handling the most toxic chemicals, highly infectious microorganisms, and certain radioactive materials. It is a 100% air exhaust hood and the exhaust is processed by filtration, incinerators, scrubbers, or similar devices designed to remove contaminants from the air stream. For certain applications the supply air to this type of hood may be HEPA-filtered as well.

All of these hoods or cabinets should be equipped with some visual indicator of proper airflow. An audible alarm also should be installed to indicate to the user whether the airflow has diminished below acceptable rates or has stopped altogether while the hood is being used and the operator is not looking at the airflow indicator (magnehelic gauge or ball flow indicator). Tissue paper flags attached to hood sashes are not reliable indicators of correct air velocity. Hoods used with highly toxic or large volumes of volatile (flammable) chemicals should be connected to emergency power should the primary electrical source fail. It is strongly recommended that they be equipped with an automatic fire extinguishing system.

All hoods should be included in a performance certification program, usually conducted by the laboratory engineering staff, the institutional safety department, or an outside consultant, in which the performance of each hood is evaluated against its design specifications for uniform airflow and face velocity across the hood opening and total exhaust air volume. The filters in Class II hoods must be checked by challenging them with a particulate smoke in order to determine whether any leakage has developed in the filter media or at seals and gaskets. Hood systems should be certified when they are installed and then at least annually unless there have been changes in the ventilation system (such as motor or fan replacements, filter changes, or the addition of other hoods to the system), the hood has been relocated, or if work operations change significantly.

## Handling and Storage of Chemicals

Operations involving the handling of chemicals, whether they be small quantities of pure or concentrated substances or large quantities of more dilute materials, provide the setting for accidental release into the environment.

It is incumbent upon the laboratory to ensure that the proper control procedures (procurement, handling, storage, and disposal), facilities, and equipment are in place and the personnel trained for the safe handling of these materials. A central repository area with controlled access is the most common method used for hazardous chemical storage.

### General Reagents (Solvents, Acids, Bases)

Safe handling begins with the procurement and receipt of hazardous materials. Common commodities such as acids, caustics, and flammable solvents should be purchased with the view that, since storage facilities in many laboratories are limited, it is preferable to order in small quantities and small container lots consistent with the needs of the operation.

### Special Reagents and Test Agents

When procuring reagent or test chemicals, it is essential that as much information as possible be obtained from the supplier or study sponsor about their physical and chemical properties, toxic effects, and procedures to follow in the event of a spill or personnel exposure. For many substances this information is available from the supplier as a material safety data sheet (MSDS) (Figure 7.3).

It should be assumed that all test chemicals received for testing are potentially hazardous, and the laboratory should establish SOPs for handling the materials during the accessioning process. It is the legal responsibility of the laboratory management to ensure that a safety program is in place and the facilities and equipment are adequate and that those who will handle the material have been properly trained to do so safely. This includes the development of written procedures and establishment not only of lines of authority and responsibility, but of accountability. Excellent resource documents are available that could be consulted in developing or improving safety programs in a variety of laboratory settings (industrial, academic, health care, and institutional).[10–12]

Storage of chemicals must be done with the same care and attention as is paid to the experimental objectives of the laboratory. Acids and bases should be stored on low shelves or in cabinets where they are not subject to falling or being easily struck. It is advisable to store glass bottles

| PRODUCT DESIGNATION |
|---|
| |

## MATERIAL SAFETY DATA SHEET

SECTION 1. SOURCE & NOMENCLATURE

| | |
|---|---|
| SPONSOR'S NAME | EMERGENCY TELEPHONE NO. |
| ADDRESS (Number, Street, City, State, Zip Code) | |
| MOLECULAR WEIGHT | CHEMICAL FAMILY |
| TEST ARTICLE NAME & SYNONYMS | FORMULA |

SECTION 2. HAZARDOUS INGREDIENTS

| BASIC MATERIAL | APPROXIMATE OR MAXIMUM % WT. OR VOL. | ESTABLISHED OSHA STANDARD | $LD_{50}/LC_{50}$ | | | |
|---|---|---|---|---|---|---|
| | | | ORAL | PERCUT. | SPECIES | CONC. |
| | | | | | | |
| | | | | | | |
| | | | | | | |
| | | | | | | |
| | | | | | | |
| | | | | | | |
| | | | | | | |
| | | | | | | |
| | | | | | | |
| | | | | | | |

SECTION 3. PHYSICAL DATA

| | |
|---|---|
| BOILING POINT (°F) | VAPOR PRESSURE (mm Hg) |
| MELTING POINT (°F) | VAPOR DENSITY (Air = 1) |
| SPECIFIC GRAVITY ($H_2O$ = 1) | EVAPORATION RATE (__________ = 1) |
| SOLUBILITY IN WATER (Pts/100 Pts $H_2O$) | VOLATILE % Vol. % Wt. |
| APPEARANCE AND ODOR | SOLVENTS |

SECTION 4. FIRE & EXPLOSION HAZARD DATA

| | | |
|---|---|---|
| FLASH POINT METHOD USED | FLAMMABLE (EXPLOSIVE) LIMITS | UPPER |
| | | LOWER |
| EXTINGUISHING MEDIA | | |
| SPECIAL FIRE FIGHTING PROCEDURES | | |
| UNUSUAL FIRE AND EXPLOSION HAZARDS | | |

| PRODUCT DESIGNATION |
|---|

SECTION 5. HEALTH HAZARD DATA (IF AVAILABLE)

| | |
|---|---|
| TOXIC LEVEL | CARCINOGENIC |
| PRIMARY ROUTES OF ABSORPTION | SKIN AND EYE IRRITATION |
| RELEVANT SYMPTOMS OF EXPOSURE | |
| EFFECTS OF CHRONIC EXPOSURE | |
| EMERGENCY AND FIRST AID PROCEDURES | |

SECTION 6. REACTIVITY DATA

| |
|---|
| CONDITIONS CONTRIBUTING TO INSTABILITY |
| CONDITIONS CONTRIBUTING TO HAZARDOUS POLYMERIZATION |
| INCOMPATIBILITY (Materials to Avoid) |
| HAZARDOUS DECOMPOSITION PRODUCTS |

SECTION 7. SPILL OR LEAK PROCEDURES

| |
|---|
| STEPS TO BE TAKEN IN CASE TEST ARTICLE IS RELEASED OR SPILLED |
| WASTE DISPOSAL METHOD |

SECTION 8. SPECIAL PROTECTION INFORMATION

| | |
|---|---|
| VENTILATION REQUIREMENTS LOCAL EXHAUST | PROTECTIVE EQUIPMENT (Specify Types) EYE |
| MECHANICAL (General) | GLOVES |
| SPECIAL | RESPIRATOR |
| OTHER PROTECTIVE EQUIPMENT | |

SECTION 9. SPECIAL PRECAUTIONS

| |
|---|
| PRECAUTIONS TO BE TAKEN IN HANDLING AND STORAGE |
| OTHER PRECAUTIONS |

| Signature | Date | Address |
|---|---|---|
| | | |

FIGURE 7.3. Material safety data sheet. All test substances should be accompanied by a safety data sheet describing any unusual chemical properties, special handling considerations, and recommended actions in the event of accidental human exposure.

in high-sided plastic trays. In the event of breakage, the contents are contained within the tray. The contents of all containers should be readily identifiable (labels intact). Care should be taken that perchloric acid, for example, is stored separately from other acids, alcohols, and organic chemicals. Ethers and dioxanes should be dated when opened and discarded as waste in 6 months because of their potential to form explosive peroxides.

Flammable liquids should be dispensed in the laboratories only in fume hoods and away from ignition sources such as burners, hot plates, or electrical equipment, and when not in use should be stored in flammable-liquid storage cabinets that comply with the NFPA No. 30, Flammable and Combustible Liquids Code.

Carcinogens (e.g., aflatoxins, nitrosamines, 3-methylcholanthrene, or dimethylsulfate) and other toxic chemicals (e.g., mitotic poisons, formalin) commonly handled in the genetics toxicology laboratory should be segregated from other chemicals in a cool, well-ventilated area or cabinet away from heat, oxidizers, or moisture. Labels must identify the contents clearly. Leaking containers should be placed in hoods or properly discarded.

Ordinary laboratory refrigerators and freezers should not be used to store volatile (flammable) chemicals. Acetone, ethers, acetonitrile, and some alcohols, for example, still generate ignitable vapors at refrigerator temperatures. A spark from the internal light switch or thermostatic control could ignite the vapors from an improperly sealed container. If flammable liquids must be kept cold, use an ice bath or purchase a laboratory-safe or explosion-safe refrigerator. These appliances have their thermostatic control relays and light switches mounted outside of the box or in a sealed unit. The considerably more expensive "explosion-proof" refrigerators are not usually required in laboratories, since they are designed for use where the ambient level of flammable vapors could routinely approach the lower flammable limit.

## PERSONNEL PROTECTION

A comprehensive safety and health program is needed that embodies the various aspects of personnel protection and a comprehensive medical surveillance program together with the use of personal protective equipment.

### Medical Surveillance Program

A comprehensive medical surveillance program embodies seven principal activities:

1. *Evaluation and maintenance of a healthful work environment.* This requires periodic inspections of work operations, environmental monitoring to determine the extent of exposure to airborne chemicals or surface contamination, and provision of health and safety data (MSDS) in safety training of employees.

2. *Preassignment examinations and/or screening.* An assessment of the employee's health status should be determined as soon as possible in order to establish baselines against which periodic health appraisals are compared. This screening may be a complete medical examination by a physician and biological testing or consist of a physician's or nurse's review of a medical questionnaire.

3. *Periodic health appraisals.* Periodic health evaluations are made in order to determine whether the individual's health remains compatible with the job assignment and to detect any evidence of occupationally related illness or injury. This may involve biological and medical monitoring. Biological monitoring involves monitoring for chemicals or metabolites that may have entered the employee's body as a result of exposure. Medical monitoring involves monitoring adverse health effects or pathologic events due to exposure.

4. *Diagnosis and treatment.* Diagnosis and treatment in occupational injury or illness cases should be prompt and aimed at rehabilitation. Physicians and nurses should be experienced in testing and diagnosing occupational exposure to a variety of chemicals.

5. *Immunization programs.* Immunization prophylaxis for employees may be indicated for certain occupations involving human pathogens or infected tissues or when working regularly with certain animals, such as subhuman primates. It is strongly recommended that clinical laboratory workers and other health care personnel be immunized against hepatitis B. A safe vaccine is now available.

6. *Medical records.* Accurate and complete medical records must be maintained for each employee.

7. *Health education and counseling.* Occupational health personnel should establish programs to educate employees in personal hygiene and health maintenance, stress and crisis management, and alcohol and drug abuse rehabilitation and treatment.

## Protective Clothing

A wide variety of personal protective gear is available, which, with proper use and care, will minimize or eliminate exposure to hazardous materials. Every laboratory worker should be familiar with the location, types, and uses of the protective gear available for the job. The laboratory supervisor should have sufficient equipment to protect everyone who will work with hazardous materials.

All laboratory workers should wear a fully-buttoned laboratory coat to protect against minor splashes or contamination of street clothes with chemicals. Even large spills will be absorbed by the lab coat and if removed promptly may not contaminate the skin. In some cases, disposable synthetic materials (e.g., Tyvek®) are used for protective clothing where maximum impermeability is desired. Tyvek is particularly applicable for work with carcinogens. Since skin contact is a major source of occupational exposure, a variety of gloves effective against rapid permeation by chemicals (acids, bases, solvents, carcinogens, etc.) should be used whenever these materials are being handled.

Eye protection should be required for all personnel and visitors where chemicals are being handled or stored. Chemical splash goggles or face shields should be worn when handling, pouring, or transferring any chemical, in order to avoid eye damage. Eye washes and drench showers should be installed where corrosive and toxic chemicals are being handled and there is risk of exposure. Contact lenses should be discouraged strongly, since gases and vapors can be concentrated under the lens and cause permanent eye damage. In the event of a splash, the involuntary eyelid spasm makes removal of the lens nearly impossible and the eye might not be irrigated satisfactorily.

Open-toed shoes, sandals, or other shoes that do not fully enclose the foot pose a significant potential for foot injuries in the laboratory. Small pieces of glass from broken pipettes or test tubes can be trapped easily between the foot and open shoe, causing serious lacerations.

## Respiratory Protection

Many laboratory operations, such as pouring, homogenization, pipetting, and centrifuging, are known to produce significant aerosols during the process. Although engineering controls (hoods, etc.) are the primary approach to containing airborne hazards, respiratory protection may also be indicated as a supplementary control measure. Respiratory protective devices vary in design, application, and protective capability; therefore, the user must evaluate the respiratory hazard and understand the particular uses and limitations of available equipment in order to ensure proper selection.

OSHA regulation 29 CFR 1910.134 obligates the employer to provide employees with respiratory protective devices when they are needed and to establish a respiratory protection program. Some of the basic elements of this program are:

1. Written standard operating procedures governing the selection and use of respirators.
2. Respirator selection based on the hazard to which the worker is exposed.
3. Instruction and training in the use and limitations of the respirator.
4. Assignment of respirators to individuals for their exclusive use, where practicable.
5. Regular cleaning and disinfection of respirators.
6. Storage of respirators in a convenient, clean, and sanitary location.
7. Inspection of routinely used respirators during cleaning; replacement of worn or deteriorated parts; inspection of emergency respirators at least once a month and after each use.
8. Maintenance of appropriate surveillance of work area conditions and degree of employee exposure and stress.
9. Regular inspection and evaluation to determine the continued effectiveness of the program.
10. Assignment of tasks requiring respirators exclusively to persons who are physically able to perform the work and use the equipment; a physician's screening of problematic health conditions.
11. Use of NIOSH/MESA-approved respirators.

Most of these requirements are straightforward and present little difficulty for implementation. However, in the toxicology laboratory, two major difficulties emerge: (1) the selection of the respirator on the basis of hazard to which the worker is exposed, and (2) the maintenance of appropriate surveillance of the degree of exposure. The major reason for these problems is that many chemicals received for toxicity testing have incomplete chemical identification and little or no physical, chemical, safety, or health data. Thus, respirator selection on the basis of hazard becomes complicated. Also, industrial hygiene air monitoring becomes impossible to perform, and therefore precludes an accurate assessment of the degree of exposure. In order to safeguard the worker from these variables, respiratory protection should be based on maximum hazard from known chemicals that have been used before in similar situations. This approach should provide an extra margin of safety.

It is also essential to assess the health status of the wearer. A complete physical examination, including a pulmonary function test, should be performed on each prospective respirator user. The results of these tests, together with the medical history, are reviewed by a physician in order to ensure that the employee is capable of wearing respiratory protective equipment.

It has been determined that persons with certain medical disorders and conditions may be at risk when wearing a respiratory protective device. Because of the added stress placed on the cardiopulmonary system, some pathologic conditions should preclude the use of respirators. Limitations on the use of these devices would be the presence of other cardiovascular or systemic diseases, which might be exacerbated.

The following clinical conditions are among those that could compromise an individual's ability to wear a respirator without risk:

1. Chronic obstructive and restrictive lung disease: chronic bronchitis, emphysema, pneumoconiosis, fibrothorax, asthma, etc.
2. Ischemic heart disease: coronary insufficiency and myocardial infarction.
3. Benign and accelerated hypertension.
4. Hemorrhagic disorders: vascular hemophilia, hypersplenism, thrombocytopenia, purpura, etc.
5. Thyroid disorders or cystic fibrosis.
6. Epilepsy: grand mal, focal, etc.
7. Diabetes mellitus.
8. Cerebrovascular accidents.
9. Facial abnormalities.
10. Kidney diseases.
11. Conductive and sensorineural hearing loss.
12. Serious defects in visual acuity.
13. Ruptured eardrum.

## WASTE DISPOSAL

Disposal of chemical wastes from the laboratory has become a complicated task in light of federal and state regulations developed over the last 5–6 years. The disposal methods available to the laboratory are incineration, removal for burial in a secure, chemical landfill, or discharge to sewers and sanitary landfills. If laboratory personnel employ good waste separation practices, the quantity of waste (solid and liquid chemicals, contaminated protective clothing, pipettes, flasks, animal bedding, paper, caging, etc.) can be kept to a minimum. In fact, a risk assessment program designed for the toxicology laboratory has been developed in order to reduce waste disposal costs, yet be consistent with environmental safety considerations.[13]

One point to bear in mind regarding toxicology laboratory waste disposal to a chemical landfill is that the reputable chemical transporter, by law, is not permitted to remove for disposal any chemical that cannot be identified by chemical name *and* physical and chemical properties. This means that test materials identified only by codes will be difficult, if not impossible, to dispose of properly unless the laboratory has incinerator or treatment facilities (both of which must be certified by the appropriate federal and/or state waste management organizations) or can obtain chemical identity information prior to disposal.

The American Chemical Society can supply to interested parties an informational pamphlet that discusses the obligations that the laboratory bears under federal waste disposal regulations (RCRA). Contact the Manager, Federal Regulatory Programs, ACS Department of Public Affairs, ACS, 1155 16th Street, N.W., Washington, D.C., 20036.

An excellent guide on laboratory waste disposal is available from the National Academy Press.[14] This resource text is a basic instructional guide and is suggested reading for anyone concerned with laboratory waste disposal.

## GOOD LABORATORY PRACTICES REQUIREMENTS

Genetic toxicology studies are often conducted as part of safety evaluation profiles for chemicals that are regulated by the U. S. Food and Drug Administration or the U. S. Environmental Protection Agency.[1,10] Both of these agencies have developed and published guidelines for good laboratory practices. The guidelines from the Food and Drug Administration became law and were effective June 20, 1979.[8] Those for the Environmental Protection Agency became effective in 1983.

The forcing factor for developing the good laboratory practices guidelines was the discovery by the Food and Drug Administration that numerous studies submitted to that agency in support of the safety of specific products could not be verified by appropriate recorded data when the studies were audited. The development of a set of good laboratory practices over the past several years has not been without controversy; even the advisability of including genetic toxicology studies under its provisions was debated. The final set of regulations, however, does include *in vitro* and other short-term testing methods.

The good laboratory practices guidelines define the conditions under which a study can be considered acceptable. The fundamental principles are that a study must be conducted in accordance with a sponsor-approved study design, there must be full documentation that each point of the study design was followed, and the results must be reported in sufficient detail to ensure that the study was conducted in compliance with the approved study design. A typical study design called for by the good laboratory practices guidelines from the Food and Drug Administration or the Environmental Protection Agency includes:

1. Name and address of the sponsor.
2. Name and address of the testing facility.
3. Professional responsible for the study (principal investigator/study director).
4. Identification of the test substance, including purity, stability, and chemical and physical properties.
5. Rationale for test organism selection.
6. Dose selection and justification.
7. Description of the test system (microorganism, cell line, or animal species).
8. Protocol to be followed, including methods of data analysis.
9. Anticipated initiation and completion dates for the study.

Once the study has been initiated, all phases of data collection must be documented by the technicians performing the tests and reviewed by the supervisor or study director. Any deviations from the detailed protocol must be described, justified, and approved by the study director. An independent quality assurance unit must monitor the ongoing study to assure compliance with the good laboratory practices guidelines. This unit also audits the raw data and final report to determine their conformity to the approved study design.

Once the study is complete a report must be prepared. The final report of a typical study will include:

1. Fully documented conditions and circumstances under which the study was conducted.
2. All pertinent data, observations, and interpretations, including data from aborted trials.
3. Description of data analysis methods.
4. Amendments and adjustments to the protocol or study with justifications and signatures of responsible professionals.
5. Dated signatures of the study director and other professional personnel involved in the study.

All raw data, preserved tissues, slides, and other specimens collected during the study must be retained in archives for a specified number of years after the data have been submitted to the regulatory agency for review. Samples of the substance tested must also be maintained for the specified period.

The requirements of the good laboratory practices guidelines are consistent with most good scientific investigations. The difference lies primarily with the principle that no piece of data is considered valid unless a complete audit of its origin can be documented. The development of the good laboratory practices guidelines has, more than anything, stimulated the development of extensive documentation and recordkeeping (Table 7.1). This requirement for documentation of all data is beginning to be felt in other areas of research and development. Some scientific journals are beginning to request a compilation of raw data with submitted manuscripts if shown in the report. This change will enable reviewers to better evaluate the quality of the studies being reported.

Good laboratory practices guidelines will not ensure high scientific quality, but they demand sufficient quality control and documentation to help assess the quality of the study. Another feature of good laboratory practices guidelines that will aid in the development of high-quality studies is the need for on-site inspection by the sponsor of a study. This forces the sponsor to interact on an intimate basis with the testing laboratory, its facilities, and the technical staff responsible for conducting the study.

The impact of good laboratory practices guidelines on a growing applied science such as genetic toxicology should aid in the standardization of methodology. Still remaining is the need for standardization in data analysis and interpretations, and for perspective related to extrapolation of test results to the human situation. Extensive work on protocol standardization is also needed in this area of toxicology. As more of this type of testing becomes part of required safety evaluation for chemicals, standardization will become increasingly important to permit objective evaluations.

No formalized guidelines are available to define protocols for all types of studies employed in genetic toxicology. Most testing laboratories rely on research publications and symposia. Thus, the variation in study designs from laboratory to laboratory is likely to be substantial. The experience developed by a laboratory, the extent of its historical data base, and the philosophy of its staff are critical factors in study design.

There has been a general concern that with the development of standardized protocols, new test modifications and research and development will be suppressed. It is unlikely that this will happen, but it would place data derived from a new or significantly modified assay into a less critical position than data derived from a recognized protocol.

The EPA Gene-Tox program described in Chapter 4 may be of greater value in defining more or less standard protocols. There are obvious advantages and disadvantages to formalizing defined protocols. The rate at which new information is being accumulated on these tests suggests that it is too early to establish highly rigid guidelines. However, the fact that toxicology tests must be conducted under good laboratory practices guidelines will at least ensure compliance with good scientific standards.

TABLE 7.1
Aspects of Studies That Must Be Verified to Meet Guidelines for Good Laboratory Practices

| Personnel | Safety | Equipment | Facilities | Records | Standard operating procedures (SOP) |
|---|---|---|---|---|---|
| Sufficient number of qualified professionals to conduct studies<br>Designated study director<br>CVs on participants of study<br>Quality assurance unit must be available to review studies | Proper storage and handling of chemicals and test compounds[a]<br>Hoods and airlocks available<br>Protective garments such as lab coats, gloves, shoe covers, and respirators available when required<br>No food, drink, or smoking in laboratory areas<br>Chemical, physical, and toxic properties as specified in safety data forms | Adequate design and sufficient quantity to conduct studies<br>Routinely cleaned and calibrated<br>Maintenance records must be available[a] | Animal areas: caging and care in compliance with Animal Welfare Act of 1970[a]; quarantine areas available<br>Waste control: sanitary disposal of animal wastes; safe disposal of hazardous wastes<br>Adequate lab space to perform studies; each chemical requires separate housing for animal studies<br>Separate administrative area<br>Data and specimen archives<br>Analysis of feed, water, and bedding[a] | Permanent record of signatures and initials of all employees[a]<br>Raw data signed and dated by technicians<br>Records of weights, calculations, measurements, and operations or observations during study[a]<br>Changes of data initialed and explained[a]<br>Data bound and stored at conclusion of study[a]<br>Historical file of each SOP[a]<br>A written protocol for each study[a]<br>Chemical identification, purity, and stability[a] | Written procedures for: animcal care; laboratory tests; receipt and identification, storage, handling, administration, and disposal of test chemical; report preparation; personnel health and safety; data handling, storage and retrieval<br>SOP must be available in laboratory at all times |

[a] Requires written documentation with names, dates, and explanations.

## ACKNOWLEDGMENT

A substantial portion of this chapter was adapted from a review by Nemchin and Brusick.[15]

## REFERENCES

1. Ehrenberg, L., and Wachtmeister, C. A.: Handling of mutagenic chemicals: Experimental safety. In *Handbook of Mutagenicity Test Procedures* B. S. Kilbey, M. Legator, W. Nichols, and C. Ramel, eds.), Elsevier/North-Holland Biomedical Press, pp. 411–418, 1977.
2. Public Law 94-580: *Resource Conservation and Recovery Act*, November 1976.
3. Office of Research Safety: Standards for Research Involving Oncogenic Viruses, DHEW Publication No. (NIH) 75-790, 3 October 1974.
4. NIH Guidelines for the Laboratory Use of Chemical Carcinogens: NIH Publication 81-2385, U.S. Department of Health and Human Services, May 1981.
5. NIH guidelines for research involving recombinant DNA molecules. *Fed. Reg.* **47**:38048–38068, 1982.
6. CDC/NIH: *Biosafety in Microbiological and Biomedical Laboratories*, HHS Publication No. (CDC) 84-8395, U. S. Department of Health and Human Services, March 1984.
7. National Toxicology Program Health and Safety Minimum Requirements for Bioassay Laboratories: In *Health and Safety for Toxicity Testing* D. B. Walters and C. W. Jameson, eds.), Baltimore Publishers, Boston, 1984, pp. 21–27.
8. National Toxicology Program Health and Safety Minimum Requirements for Cellular and Genetic Toxicology Laboratories: In *Health and Safety for Toxicity Testing* (D. B. Walters and C. W. Jameson, eds.), Baltimore Publishers, Boston, 1984, pp. 28–30.
9. Department of Health, Education and Welfare, Food and Drug Administration: Good laboratory practices regulations for nonclinical laboratory studies. *Fed. Reg.* **43**(Part II):59986–60020, 1978.
10. Steere, N. V. (ed.): *Handbook of Laboratory Safety*, 2nd ed., Chemical Rubber Co., Cleveland, 1971.
11. Manufacturing Chemists Association: *Guide for Safety in the Chemical Laboratory*, 2nd ed., Van Nostrand Reinhold, New York, 1972.
12. National Research Council, Committee on Hazardous Substances in the Laboratory: *Prudent Practices for Handling Hazardous Chemicals in Laboratories*, National Academy Press, Washington, D. C., pp. 193–212, 1980.
13. Nemchin, R. G., and Coco, J. A.: A risk assessment program for toxicology laboratory waste disposal. In *Health and Safety for Toxicity Testing* (D. B. Walters and C. W. Jameson, eds.), Butterworth Publishers, Boston, pp. 283–297, 1984.
14. National Research Council, Committee on Hazardous Substances in the Laboratory: *Prudent Practices for Disposal of Chemicals from Laboratories*, National Academy Press, Washington, D. C., 1983.
15. Nemchin, R. G., and Brusick, D. J.: Basic principles of laboratory safety. *Environ. Mutagen.* **1**:947–971, 1985.

CHAPTER 8

# Descriptions and Evaluation of Genetic Toxicology Assays

## GENERAL CLASSIFICATION OF GENETIC TOXICOLOGY ASSAYS

A significant level of redundancy exists in the types of tests included in genetic toxicology. In an effort to group these tests into similar classes of tests, four groups have been defined (Table 8.1). Two groups of tests measure genetic changes in the form of specific locus-gene mutation and chromosomal aberrations. A third group, primary DNA damage tests, typically measure the induction of DNA adducts or breakage and the cellular responses produced during the repair of this damage. These responses include stimulation of DNA repair, DNA–adduct formation, somatic recombination between homologous or sister chromatids, or DNA strand breakage. A fourth group does not directly identify DNA damage, but assesses the ability of a test substance to transform "normal" mammalian cells into cells with neoplastic properties. It has been assumed that DNA alterations are part of the cell transformation process and more recently the analysis of mutational activation of transforming oncogenes, such as the *ras* group, had added additional evidence to that assumption.[1] Primary DNA damage tests are used predominantly as a measure of oncongenic potential and not associated with the production of transmissible genetic damage.

## LEVEL OF ASSAY DEVELOPMENT

While many test methods are available within each of the four classes identified above, selection among those available for inclusion into a test

TABLE 8.1
Classification of Short-Term Assays

| Group | Characteristic of events detected |
|---|---|
| Gene mutation | Intragenic nucleotide changes in leading to substitution or addition/deletion |
| Chromosome aberrations | Structural and/or numerical changes in normal karyotype |
| Primary DNA damage | Heterogeneous group of DNA interactions including DNA adducts, strand breaks, stimulation of DNA repair, SCE |
| Morphologic transformation | Morphologic changes in target cells associated with tumorigenicity |

battery is often done without essential information about the performance of the particular assay. Tests are often employed because they happen to be recommended by the testing laboratory or are based upon recommendations from colleagues or recently read journal articles. This approach will not necessarily result in the most appropriate set of tests for use in a given situation. One wants to select tests that have a good track record and are also known to function reliably and accurately most of the time (see Chapter 4).

The following descriptions of test systems also includes some of the advantages and limitations associated with use of the assays in safety testing.

## DESCRIPTION OF COMMON ASSAYS FOR GENE MUTATION

### Microbial

There are large numbers of point mutation assays in bacteria, yeast, and mold. Each type of organism has its own specific attributes and limitations. Mutation assays within the three phylogenetic categories of microorganisms have been reviewed in detail.[2,3] All have been subjected to a diverse group of chemical mutagens, they are generally reproducible, detailed protocols for each have been published, and all can be used with an S9 activation system.

1. *Bacteria*. The Ames *Salmonella*/microsome assay for reverse mutation measures mutation in a series of histidine-requiring auxotrophs of *S. typhimurium*. Mutant strains represent "hot spots" and detect both base-pair substitution and frameshift mutagens. The basic series of strains routinely employed has undergone several modifications to increase their

TABLE 8.2
Genotype of the TA Strains Used for Testing Mutagens[a]

| Additional mutations in: LPS | Repair | Contains plasmid | Histidine mutation in strain hisG46 | hisC207 | hisC3076 | hisD3052 |
|---|---|---|---|---|---|---|
| + | + | — | hisG46 | hisC207 | hisC3076 | hisD3052 |
| + | Δ*uvrB* | — | TA1950 | TA1951 | TA1952 | TA1534 |
| Δ*gal* | Δ*uvrB* | — | TA1530 | TA1531 | TA1532 | TA1964 |
| *rfa* | Δ*uvrB* | — | TA1535 | TA1536 | TA1537 | TA1538 |
| *rfa* | + | — | TA1975 | TA1976 | TA1977 | TA1978 |
| *rfa* | Δ*uvrB* | pKM101 | TA100 | — | TA97 | TA98 |
| *rfa* | + | pKM101 | — | — | — | TA94 |
| + | Δ*uvrB* | pKM101 | TA2410 | — | — | — |
| + | + | pKM101 | TA92 | — | — | TA2420 |

[a] From Ames *et al.*[4] All strains were originally derived from *S. typhimurium* LT-2. Wild-type genes are indicated by +. The deletion (Δ) through *uvrB* also includes the nitrate reductase (*chl*) and biotin (*bio*) genes. The Δ*gal* strains (and the *rfa, uvrB* strains) have a single deletion through *gal chl bio uvrB*. The *rfa*, repair+ strains (last row) have a mutation in *galE*. Strains containing pKM101 carry an extra chromosomal element (plasmid), which enhances error-prone repair.

sensitivity. The origin and development of these strains is given in Table 8.2. Maron and Ames revised some of the methods for the *Salmonella* assay.[5] In that publication two new tester strains were proposed as possible additions to and/or replacements for some of the existing strains.

Strains TA1535 and TA100 were derived from hisG46 and contain -G-G-G- at the mutant site. Consequently, chemicals with an affinity for A:T base pairs might not revert this strain. One of the new strains, TA102, was derived from hisG428 (not shown in Table 8.2) and contains -T-A-A- at the mutant site.[6] It also has multiple copies of the mutant site on plasmid DNA to increase target site numbers per cell. Use of both TA100 and TA102 should be complementary and permit detection of both G:C- and A:T-specific chemicals.

Strain TA97, derived from hisC3076, has been recommended as a replacement for strain TA1537. This strain was designed to detect similar mutagens as TA1537, but should be somewhat more sensitive due to the R-factor and to a hot spot, which broadens its response to a greater array of frameshift mutagens.

Most other point mutation assays that measure either forward or reverse mutation in bacteria duplicate the detection pattern of the Ames assay and can be expected to add little to the results obtained with *Salmonella*.[7,8]

2. *Yeast*. Both forward and reverse mutations can be detected by yeast assays. Forward mutation to canavanine resistance is a convenient

TABLE 8.3
A Comparison of Screening Techniques Employing Yeasts

| Test | Organism[a] | Haploid (H) or diploid (D) | Strain | Genetic damage detected |
|---|---|---|---|---|
| Mutations | S.c. | H | S288C | Forward mutation from canavanine[s] to canavanine[r] |
| | S.p. | H | P1 | Forward mutation at five genes preceding all *ade*-6 in purine biosynthesis |
| | S.c. | H | S138/S211 | Reverse mutation in two *met*[−] auxotrophic strains |
| | S.c. | H | XV185-14C | Reverse mutation at *his*-1, *hom*-3, and *arg*-4 loci |

[a] S.c., *Saccharomyces cerevisiae*. S.p., *Schizosaccharomyces pombe*.

and reliable test.[9] Canavanine is a toxic analog of arginine, and when placed in growth medium of normal haploid yeast [canavanine-sensitive (*can*$^{s}$)], uptake of canavanine by the arginine transport kills the *can*$^{s}$ cells. Cells that have undergone a mutation in the transport gene, rendering it inactive, synthesize arginine *de novo* and are resistant to the canavanine (*can*$^{r}$) in growth medium. This method is capable of detecting both base-pair substitution and frameshift mutagens.

Yeast assays have a general limitation in that the cell membrane restricts the permeability of certain molecules that might have genotoxic properties but cannot reach the target DNA. Numerous studies have reported significant microsomal metabolism by growing yeast cultures.[10] This might be considered either an advantage or a disadvantage, since activation of chemicals is often required; however, target cell metabolism is generally not desirable, since there is a chance that it will form unique metabolites not generally produced by mammalian microsomal enzymes. Most yeast strains, whether +S9 or −S9 mix, are currently conducted using growing cells and with extended exposures of 18–24 hr. Reverse mutation studies, either quantitative (most commonly employed) or as plate incorporation methods, are useful tests in specific instances. Several strains have been constructed to evaluate reverse mutation in yeast (Table 8.3).

## Mammalian Cells *in Vitro*

Table 8.4 lists several of the common gene mutation tests conducted in cultured mammalian cells. Most of these tests utilize one of three loci, thymidine kinase (TK), hypoxanthine-guanine phosphoribosyl transferase

TABLE 8.4
Some Commonly Used *in Vitro* Mammalian Cell Lines for Gene Mutation Studies

| Cell line | Gene or trait used for mutant selection[a] |
|---|---|
| Mouse lymphoma L5178Y | TK, HGPRT, OR |
| CHO | HGPRT, OR |
| V79 hamster cells | HGPRT, OR |

[a] TK, thymidine kinase, autosomal recessive. HGPRT, hypoxanthine-guanine phosphoribosyl transferase, X-linked gene. OR, ouabain resistance.

(HGPRT), and resistance to ouabain. The biochemical steps associated with the use of each are given in Figure 8.1. The HGPRT and TK loci should be capable of detecting both base-pair substitution and frameshift mutagens, whereas the ouabain gene locus will probably be suboptimal for detecting frameshift mutagens. The reason is that frameshift mutagens often result in premature termination of gene transcription, with partial or complete loss of gene product. Expression of ouabain resistance requires the presence of a complete protein; thus, frameshift mutations would be lethal to the cell rather than mutagenic.

All of the techniques listed in Table 8.4 are performed with mammalian S9 preparations similar to those used in the Ames *Salmonella* assay.

Due to the many aspects of performance (suspension exposure, agar cloning, short expression time), the mouse lymphoma Assay is believed to be somewhat more sensitive than the V79 or CHO cell assays. The L5178Y cell line is also more responsive due to the fact that presence of the TK gene locus is an autosomal recessive trait. Consequently, large chromosome deletions induced in the normal allele of the TK+/− heterzygote that cover more than the TK gene will not be lost as lethals, since the corresponding alleles on the homologous chromosome will provide the necessary functions. Large deletions in the hemizygous HGPRT locus of CHO and V79 cell lines are likely to be expressed as lethality, thus reducing the responsiveness in these assays toward clastogenic agents.

Recent modifications in the CHO/HGPRT protocol assay that allow for suspension exposures have resulted in greater sensitivity. Tindall *et al.* recently introduced a new clone of CHO cells that has the normal HGPRT locus deleted and a new HGPRT engineered into the cell from bacterial origin.[11] The insertion site of the new gene locus appears to be a place where its deletion does not result in lethality, but rather in a

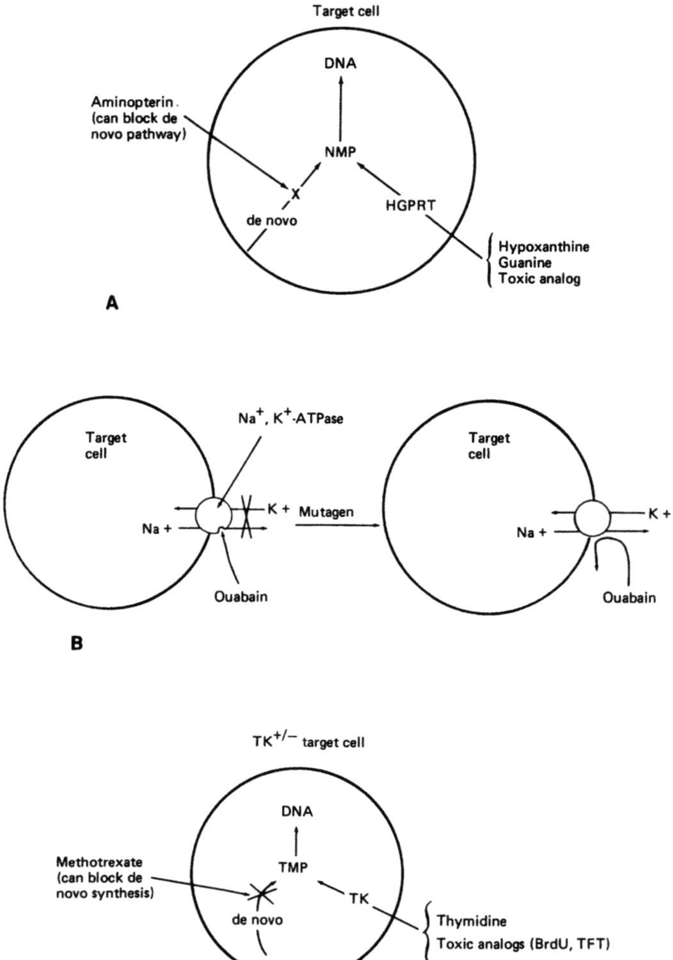


FIGURE 8.1 (A) Selection of mutagen-induced HGPRT$^+$ phenotype. A mutation in the gene makes the target cell insensitive to the toxic selective agents, and DNA is synthesized *de novo*. Azaguanine and thioguanine are two analogs used to select for HGPRT mutants. NMP, Nucleoside monophosphate. (B) Selection of mutagen-induced ouabain-resistant (Oua$^r$) phenotypes. Ouabain affects the $Na^{2+}$ and $K^{2+}$ transport enzyme system. Mutant cells have an altered protein which is no longer insensitive to ouabain, yet permits normal $Na^{2+}$ and $K^{2+}$ transport. Complete (nonleaky) mutants will probably be lethal and not detected in this assay. (C) Selection of mutagen-induced TK−/− phenotypes. Mutant TK−/− cells are resistant to the pyrimidine analogs BrdU or trifluorothymidine (TFT) and synthesize their DNA *de novo*. Nonmutant cells (TK+/−) will die in the presence of BrdU or TFT.

recoverable mutant. This cell line increases the susceptibility of the CHO clone to clastogens.

For general chemical screening, the mouse lymphoma assay controlled for deviations in pH and osmolality appears to be the most useful technique.

## Insects

The only routine gene mutation assay in insects is the sex-linked recessive lethal (SLRL) assay in *Drosophila melanogaster*. The utilization of this technique in genetic toxicology has been the subject of more than one review.[12–14] All classes of point mutations can be detected in germ cells with the *Drosophila* test, and this test has even been suggested for use in risk estimations for heritable effects. Its major limitation is associated with the difficulties often encountered in quantitatively determining the dose actually administered to the flies by feeding. As a result, extensive preliminary toxicity studies to ascertain compound uptake are often required prior to testing to ensure proper exposure, especially with negative studies. Microinjection techniques are possible and can overcome problems of test material uptake.

Sampling of different stages of spermatogenesis in *Drosophila* has demonstrated germ-cell-stage specificity in response to mutagens, and the type of microsomal metabolism and detoxification found in both mice and flies is sufficiently close to suggest to some that *D. melanogaster* be given the status of "Honorary Mammal."[15]

While the basic methodology of the SLRL is rather standardized, *Drosophila* strains can vary from laboratory to laboratory. Recently, repair-deficient strains have been isolated, and are under analysis to determine if they will have advantages over the traditional strains.[16] None of the strains currently evaluated appears to be adequate to replace traditional stock cultures.

A somatic mutation test has also been developed that detects somatic recombination, mutation, and chromosome deletions.[17] The technique employs changes in wing hairs to identify the above events. Wings of flies treated as larvae are examined microscopically for evidence of changes in wing hair morphology. Mutagens, such as direct-acting alkylating agents and nitrosamines, and diverse agents, including 5-aminoacridine, hydroxylamine, and aflatoxin B1, are active in this test. The efficiency of this test when compared to the SLRL test is quite dramatic, since positive effects can be observed with as few as 50 individual wings.

The *Drosophila* SLRL assay is not appropriate as a screening test and is applied only in specific circumstances. Somatic cell mutation assays should probably be used in place of the SLRL assay when possible.

TABLE 8.5
Animal Methods Employed for Genetic Risk Analyses

| Bioassay | Technique | Reference |
|---|---|---|
| Specific-locus test in mice | Measurement of visible atuosomal recessive mutation in mice using a multiply marked tester strain of mice containing up to seven recessive loci; minimal size of experiment needed to define negative would be in the range of 12,000–14,000 offspring of treated animals | 18, 19 |
| Cataract mutation assay in mice | Measurement of visible dominant mutations in mice using a biomicroscopic examination of the eye with a slit lamp; $F_1$ offspring of treated mice are examined at 3 weeks of age; no special strains of mice are required; this assay can be used to make direct calculations of induced mutation rats | 21, 22 |
| Skeletal mutation assay in mice | Measurement of visible dominant mutations affecting the skeletal system of mice; requires careful analysis of bone development | 23 |

## Mammals

Gene mutation assays using mammals has been the domain of those performing risk analysis. The methods available include recessive and dominant morphologic traits and recessive biochemical traits. Table 8.5 summarizes the most common of these germ cell tests.

The mouse specific-locus test has been used for over 20 years to determine the mutagenic risk of chemicals.[19] Few laboratories are set up to conduct the mouse specific-locus assay, because of the animal breeding requirements necessary to perform a study. Dominant mutation assays tend to require less in the way of resources, but are often limited to a few laboratories and are generally investigator-specific. Most of the *in vivo* germ cell assays have not been subjected to a broad range of chemicals or chemical classes.

Assays for *in vivo* somatic cell mutation are also limited. The most common test is the mouse somatic-cell coat-color mutation assay (spot test).[20] The target cells are embryonic melanocytes, and the potential for placental inhibition of chemical distribution and the fact that the test can only be conducted in pregnant female mice (hormonal influences) detract from the utility of this method. Even with these disadvantages, the test has a good track record and appears to respond to many classes of chemical mutagens.[24] Among 45 carcinogens tested, 35 (78%) were positive in the spot test. The mouse spot test is probably the most useful and reliable method to investigate single-gene mutagenesis in intact mammals.

## TESTS FOR CHROMOSOME ABERRATIONS

### Microbial

Most microbial assays employ haploid indicator organisms, and are not capable of demonstrating chromosome damage. Chromosome breakage in these cells results in lethality. Some fungi, however, can be used to study specific types of alterations involving chromosomes. *Neurospora crassa* can detect multilocus chromosome deletions in the *ade*3 heterokaryotic system[2] and chromosomal nondisjunction can be measured in *Aspergillus nidulans*[25] and yeast.[26] These microbial assays are not routinely employed in genetic toxicologic studies, because of the organisms' distant phylogenetic relationship to animal systems and the requirements for highly specialized equipment and facilities.

Diploid yeasts have also been proposed as models for aneuploidy assessment. Some rather convenient techniques using *Saccharomyces cerevisiae* are available, but one cannot be sure that chromosome loss (marker loss) in yeast is mechanistically related to mammalian aneuploidy, especially when one is somatic (yeast) and the other meiotic (mammalian). When considering chromosomal modifications, either structural or numerical, cell cycle mechanisms and chromosome movement might be extremely important in generating this type of genetic damage. These processes appear to be different between fungi and animals, making extrapolation of fungal results to mammals difficult. Consequently, microbial models for aneuploidy are probably dead ends as safety assessment tools.

### Mammalian Cells *in Vitro*

Almost any cell line with a well-defined karyotype could be used for *in vitro* cytogenetic analysis. Primary considerations are growth potential, chromosome number, and chromosome diversity. Rat chromosomes, for example, are almost entirely acrocentric, somewhat increasing the difficulty in scoring. CHO cells have a low number of chromosomes, which are diverse in appearance, making scoring less tedious; they are therefore preferred by many laboratories for routine analyses. Studies with cell lines can be conducted with S9 microsome activation systems. Cultured human lymphocytes are often used because of their perceived relevance, but there seems to be little if any data to support this perception. It should be stressed that *in vitro* methods for cytogenetic analysis can be misinterpreted. High concentrations of nontoxic materials may change pH or osmotic conditions to an extent that chromosome breakage occurs. Consequently, pH and osmotic controls (or limits) should be incorporated into all *in vitro* tests for aberration analysis.

*In vitro* chromosome aberration analysis plus an Ames test appears to be a generally accepted minimal screening battery for new chemicals being introduced into the environment. Sufficient data are not yet available to document the accuracy of this two-test battery.

## Insect Tests for Chromosome Effects

*Drosophila melanogaster* is a versatile test organism, which can be used to detect heritable chromosome effects, including sex-chromosome (X or Y) loss and heritable translocations. Again, both effects are measured in germ cells exposed to the test substance.

The test for loss of X and Y chromosomes in *D. melanogaster* is a rapid, one-generation screen for chemicals that produce chromosome breakage or nondisjunction. In *Drosophila,* the sex of the individual is determined by the ratio of X to Y chromosomes. The possible viable combinations resulting from normal and abnormal segregation are:

| Configuration | Expression |
|---|---|
| XY | Normal male |
| XX | Normal female |
| XO | Sterile male |
| XXY | Superfemale (abnormal but functionally unaffected) |

The chromosomally unbalanced individuals, XO and XXY, are relatively unaffected because the X and Y are structurally very unlike one another; the X contains a great deal of information and the Y very little. The X is estimated to contain 20% of the total genome of the fly, while the Y contains presumably very little other than male fertility factors. Only the configurations Y and XXX are inviable. By marking the sex chromosomes with genes for readily distinguishable phenotypes, it is easy to track the progress of the treated chromosomes as they are passed to the offspring.

Reciprocal, balanced translocations can be readily measured in *Drosophila* by "engineering" stocks that have a visible mutant on their major autosomes (non-sex chromosomes). The simplest markers to use are those for eye color, since the technician can then look at only one localized body region when scoring for mutants. The genetic principle or rationale for the test is that each autosome has a homolog or "mate" that provides a similar amount and type of information. The information provided may of course be slightly different at each locus, but for the individual to survive, large portions of the chromosome can neither be deficient nor occur in excess. Two or more breaks induced in a chromosome that then rejoin improperly, transferring parts of chromosomes, will result in viable offspring only if there is an equal or reciprocal exchange of information (a translocation).

Reciprocal translocations induced in wild-type males exposed to chemicals can be detected by crossing the treated male to a female homozygous for visible mutants on the second and third chromosomes, backcrossing the sons of this cross to a female identical to the mother, and examining the $F_2$ male progeny for the presence of flies with four eye colors (red, white, orange, and brown). Two of the eye-color classes resulting from such a male (orange and brown eyes) indicate an unbalanced amount of information.

Both of these endpoints measure chromosome alterations analogous to sex-chromosome and translocation events in mice. Thus, if one assumes sufficient similarity in gametogenesis and metabolism, the *Drosophila* system may function as a model of the whole animal in measuring for heritable chromosomal alterations.

## Mammals

Rodents have long been utilized in genetic toxicology for evaluating chromosomal alterations. Most of the work has been conducted in mice, with fewer studies using rats, hamsters, and assorted other small rodents. The *in vivo* bone marrow analysis for chromosomal effects, the bone marrow micronucleus assay, and the dominant lethal test have respectable data bases in both mice and rats. Heritable translocation analyses and sex-chromosome loss are routinely conducted in mice.

### Rodent Chromosome Analysis

One of the major considerations involved with rodent chromosome analysis, once the target site and species have been selected, is proper design for cell cycling sampling (hours postexposure for cell harvest), number of animals and/or cells to score for aberrations or micronuclei, and dose selection. Sampling times and dose selection can best be determined from data on mitotic suppression developed from preliminary dose range studies. The sample sizes for analysis require statistical analysis based on the level of sensitivity desired in the specific study.

TABLE 8.6
The Major Differences between the Standard Dominant Lethal Protocol, the Extended Dosing Protocol, and the European Working Group Protocol[27]

| | Standard | Extended dosing | Maximized |
|---|---|---|---|
| Animals per dose level[a] | 15–20 | 15–20 | 50 |
| Dosing | 5 days | 8–10 weeks | 5 days |
| Mating periods (number × duration) | 7 × 5 days | 2 × 5 days | 13 × 4 days |
| Rest between mating periods | 2 days | 2 days | — |
| Females per male during mating | 2 | 2 | 1 |
| Total number of animals requires for a protocol including three dose levels and positive and negative control groups: | | | |
| Male | 75–100 | 75–100 | 250 |
| Female | 1050–1400 | 300–400 | 3250 |

[a] Better data can be obtained if proven breeder males are used.

## Dominant Lethal Assays

Tests for dominant lethality have been conducted since 1953. They typically provide information on the fate of the zygotes produced from the sperm of treated male animals and the eggs of untreated females. The test measures death that occurs between fertilization and parturition. Events such as failure of the zygote to implant in the uterine wall or to survive to midpregnancy after implantation are detected upon necropsy of the mated females. It is generally assumed that the vast majority of such dominant lethal effects are the result of chromosome alterations, either structural or numerical. Cytogenetic analysis of preimplanted zygotes for chromosomal integrity can be performed to confirm this mechanism. This technique is extremely tedious, since it requires recovery of fertilized eggs from the fallopian tubes and uteri; however, analyses have clearly established that structural and, especially, numerical chromosome alterations are responsible for most of the observed dominant lethality in rodents.

There are three rather standardized approaches to an evaluation for dominant lethality. The primary differences are shown in Table 8.6. The mouse and rat are both acceptable species; selection of species is often made to coincide with other toxicologic data.

The traditional approach to an evaluation for dominant lethality consists of an acute or subchronic exposure followed by sequential weekly matings for 7–10 weeks. Each week, the dosed males are caged with two

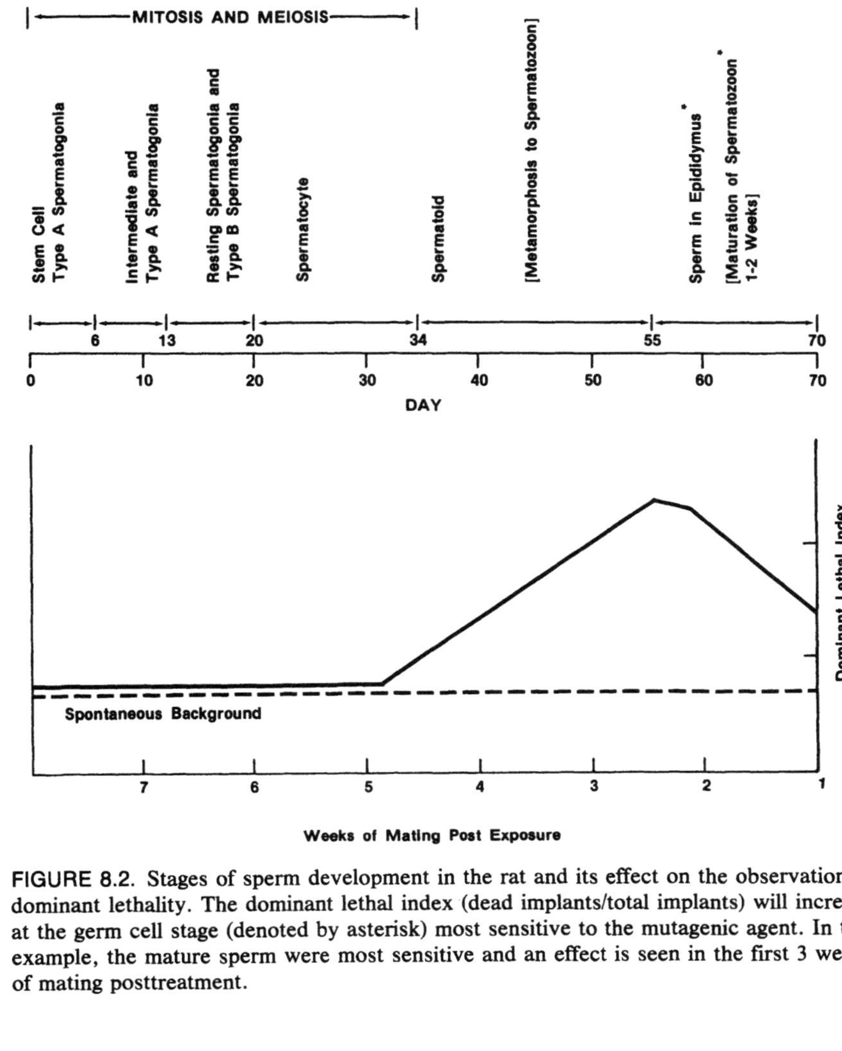


FIGURE 8.2. Stages of sperm development in the rat and its effect on the observation of dominant lethality. The dominant lethal index (dead implants/total implants) will increase at the germ cell stage (denoted by asterisk) most sensitive to the mutagenic agent. In this example, the mature sperm were most sensitive and an effect is seen in the first 3 weeks of mating posttreatment.

virgin females. The females are sacrificed at midpregnancy and necropsied. By 7 weeks postexposure, all cell stages that were present at the time of exposure to the test material have traversed the entire meiotic cycle and been used to fertilize the females. If the test material has a specific effect on one or more cell stages, the effect would be observed as an elevation in dominant lethal parameters at the time of most susceptibility (Figure 8.2).

The second approach follows a scheme where the males are dosed for 7–10 weeks and then mated twice. This extended dosing method permits all cell stages to have continuous exposure during one round of meiosis. All damage will accumulate in the sperm and presumably be expressed in either or both of the two matings. The advantages associated with this method are related to the fewer numbers of animals involved in conducting two matings rather than seven or ten. The accumulation of dominant lethality in the two matings may produce an enhanced response if the test substance is genotoxic.

A maximized protocol has been recommended, especially by European investigators.[27] This study design increases the number of dosed males to 50 per dose group. The dosed males are subjected to sequential mating; however, the matings are made 1:1, male to female, and carried out on 13 consecutive 4-day cagings. The 4-day cagings are based on the estrous cycle of the mouse and the assumption that if mating occurs, it will occur during the first 3 days of caging male and female mice. This procedure is considerably more extensive than the two previously described, but provides considerably more information. Using this approach, small increases over the spontaneous level can be detected with a high degree of confidence. This version of the test, however, requires 3500 animals when positive and negative controls and three concentrations of the test material are included in the study design.

Dominant lethal assays can be performed on both male and female mammals, but studies in females are not performed routinely, because oocytes are generally less sensitive to chemicals than are sperm. It is difficult, therefore, to unequivocally discriminate between cytotoxic and true genetic effects.

The high spontaneous background of dominant lethality in rodents (6–7%) and the inability to verify heritability makes the dominant lethal assay a suboptimal method for sensitive estimations of germ cell risk unless large numbers of animals are used. Since almost all dominant lethal effects are due to chromosomal alterations, it appears that *in vivo* cytogenetic analysis (bone marrow cells) would be a more sensitive screen for clastogenic agents, thus limiting the need for dominant lethal assays for compounds that fail to show somatic clastogentic activity. Holden[28] analyzed the dominant lethal data base and was able to demonstrate that positive compounds were all detectable by bone marrow aberration studies.

### Heritable Translocation Assay

This assay measures chromosomal translocations produced in germinal tissue that are capable of being transmitted to the next ($F_1$) generation. Translocation heterozygotes are characterized by reduced fertility (partial or complete); detection of the heterozygotes is made on the

Functions

| WEEK | 1 | 2 | 3 | 4 | 5 | 6 | 7 | 8 | 9 | 10 | 11 | 12 | 13 | 14 | 15 | 16 | 17 | 18 | 19 | 20 | 21 | 22 |
|---|---|---|---|---|---|---|---|---|---|---|---|---|---|---|---|---|---|---|---|---|---|---|
| Function | Quarantine | | Dose Compound | | | | | | | Mate $F_0$ | Gestation and Delivery | | | | Wean $F_1$ Males | | | | Select Groups of $F_1$ Males | | | |

Functions

| WEEK | 23 | 24 | 25 | 26 | 27 | 28 | 29 | 30 | 31 | 32 | 33 | 34 | 35 | 36 | 37 | 38 | 39 | 40 | 41 | 42 | 43 | 44 |
|---|---|---|---|---|---|---|---|---|---|---|---|---|---|---|---|---|---|---|---|---|---|---|
| Function | Mate $F_1$ | Gestation and Kill | | Submit Data | | Review of Data | | | | Cytogenetic Analysis of Presumptive Translocation Heterozygotes | | | | | | | | | | | Final Report | |

Review Data for Selection of Presumptive Carriers

Kill Animals/Prepare Slides

FIGURE 8.3. Typical performance schedule for the heritable translocation assay.

TABLE 8.7
Analysis of Heritable Translocation Assay: Definition of the Sensitivity of the Test[a]

| Number of $F_1$ males mated | Induced (true) translocation rate (%) | $p$ value[b] | Level of significance ($\alpha$)[c] |
|---|---|---|---|
| 100 | 1.0 | 0.26 | $\alpha \leq 0.01$ |
| | 1.5 | 0.44 | ($C \geq 2$) |
| | 2.0 | 0.60 | |
| 300 | 1.0 | 0.80 | $\alpha = 0.03$ |
| | 1.5 | 0.94 | ($C \geq 2$) |
| | 2.0 | 0.98 | |
| 600 | 1.0 | 0.94 | $\alpha = 0.02$ |
| | 1.5 | 0.99 | ($C \geq 3$) |
| | 2.0 | >0.99 | |

[a] From Generoso *et al.*[29] This table defines the likelihood of detecting a given rate at $\alpha$ for a given population of $F_1$ males. Thus, if one is looking for induced rates of 1–2%, a group of 100 animals is suboptimal; a level of 300 $F_1$ males gives a much more accurate assessment.
[b] Assuming the spontaneous translocation carrier frequency is $9.1 \times 10^{-4}$ in untreated $F_1$ males, this gives the chance of detecting the rate given in column 2.
[c] $C$ is the minimum number of translocation carriers that need to be observed in order to conclude that the measured rate is significantly greater than the spontaneous rate with a probability of $\alpha$.

basis of litter size (living and dead embryos) after mating the dosed males with normal females. Confirmation is generally assessed by cytogenetic analysis of gonadal cells (spermatocyte evaluation) for recognizable meiotic chromosome figures indicative of translocations. Although direct analysis for translocation-bearing spermatocytes in $F_1$ progeny can be performed without the preliminary mating sequences, preselection of presumptive translocation carriers via fertility testing is the standard approach, since it reduces the number of animals that have to be cytologically analyzed.

The test uses the mouse as the target species because of the ability to clearly detect chromosome translocation figures in spermatocytes of this species, and because of cost. A typical translocation experiment may include anywhere from 4000 to 10,000 animals, depending on the level of sensitivity desired.

There is a certain analogy between the dominant lethal and heritable translocation assays, since both tests measure chromosomal damage. Only specific types of chromosomal damage, such as reciprocal translocations, are compatible with cell viability and will be transmitted to the $F_1$ generation. This is expressed by the very low (0.1%) frequency of translocation heterozygotes found in control populations of mice.[29]

A general outline of a mouse heritable translocation test is shown in

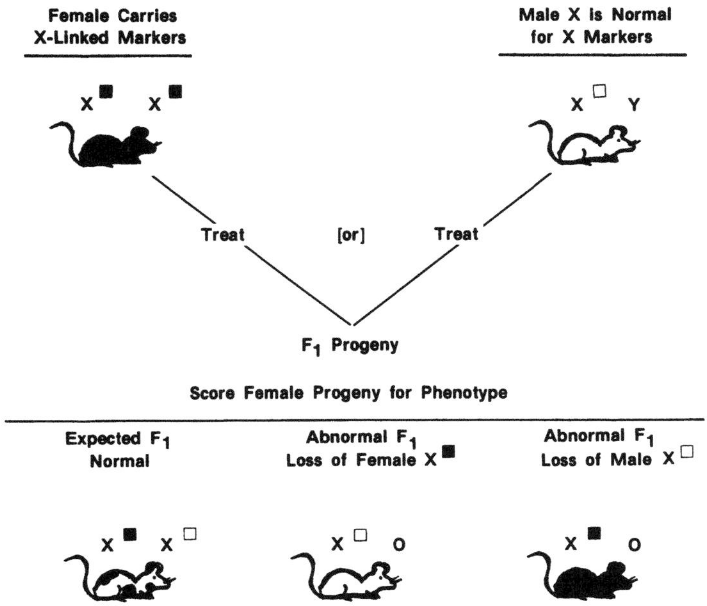


FIGURE 8.4. Sex-chromosome loss in mice. This assay is based on the fact that XO female mice are viable. Distinct genetic traits placed on the two chromosomes are used to show which X chromosome is lost. The normal XX female shows both traits.

Figure 8.3. A test requires between 30 and 35 weeks, depending upon the number of matings to be conducted. Table 8.7 is useful in developing protocols, since it defines the likelihood of detecting a given translocation rate α for a given population of $F_1$ males sampled.[29] For example, 300 $F_1$ males are necessary to detect an induced rate of 1–2% with a level of confidence near 95%. This depends somewhat on the criteria used to define $F_1$ males as fertile. A good analysis of the effect of criteria on the number of animals required for heritable translocation evaluations in mice is given by Wiemann and Lang.[30]

Both Generoso *et al.*[29] and Weimann and Lang[30] reported that translocation carriers induced at a rate of 1–2% can be detected with a relatively high degree of confidence using reasonable numbers of animals. For example, a test with a negative control group and a single dose level might use about 5000 animals. Dose–response evaluation, however, would involve significant time and expense.

## Sex-Chromosome Loss

The test for sex-chromosome loss in mice is not commonly conducted in genetic toxicology laboratories. Like the *Drosophila* assay, it is based upon phenotypic genetic markers located on the X chromosome. If either sex incurs an X loss, the $F_1$ female progeny resultant from the loss can be identified (Figure 8.4). Sex chromosome nondisjunction is presumably detected with this approach. Nondisjunction accounts for the greatest proportion of human disorders associated with chromosomal alterations, but there is little evidence in mammals that chemical treatment is responsible for a significant portion of nondisjunction.

## TESTS FOR PRIMARY DNA DAMAGE

This group contains a heterogeneous array of tests. Results from these tests are difficult to fit into a hazard assessment if not supported by data from gene mutation or chromosome assays. Some of the tests, such as those detecting somatic recombination and DNA strand breakage, measure direct genotoxic effects that would be expected to produce mutation or cell death. Others, such as the tests for DNA adducts, UDS, and SCE, are less direct indicators of genetic toxicity. For example, the process of UDS is not normally considered a mechanism for genetic damage; rather, it is the expression of DNA damage repair, which should reduce the level of mutation or chromosomal breakage. The process of reciprocal exchange between identical sister chromatides in the SCE assay should not, theoretically, lead to genetic damage; however, some investigators have suggested a general relationship between SCE induction and mutation induction.[31] Similar arguments are used to justify adduct analysis.

The application of primary DNA damage tests to genetic testing is currently the subject of debate. One argument proposes that gene mutation and chromosome aberration tests should be the only ones relied upon for the assessment of mutagenic hazard. Tests for primary DNA damage, although contributing to positive findings in gene mutation or chromosome aberration assays, cannot be viewed, alone, as indications of potential human health hazard. It would be difficult to presume significant genetic hazard from a chemical that does not induce point mutations or chromosome aberrations, but does induce SCE.

Similar to other endpoints, those encompassed by the general category of primary DNA damage are detected in a variety of submammalian, *in vitro,* and mammalian species.

### Microbial Tests

#### Bacteria

Several genera of bacteria have been employed as indicators of primary DNA damage. The basis for microbial tests rests on the ability to derive repair-deficient (usually excision) substrains from normal repair-competent strains. The two bacterial substrains should be identical in all respects except for their ability to repair DNA lesions. If the two substrains are exposed to a genotoxic material, the nonrepaired lesions in the deficient strain will result in lethality, whereas the same concentration of the genotoxic material in the normal strain is less toxic. Nongenotoxic chemicals should exhibit roughly equivalent lethality.

The effect of differential lethality is not a measure of hereditary changes, but only identifies agents that impact on DNA, resulting in lethal damage. The damage may or may not be of a type that produces hereditary effects.

Several bacteria systems, including those using *E. coli,*[32] *S. typhimurium,*[4] and *B. subtilis,*[33] are routinely employed in these tests; and interpretation of effects is based on the ratio of lethality in the two strains. The results of this group of tests are of little value making in safety testing assessments.

### Yeast

The processes of somatic recombination are included in the group of tests that detect primary DNA damage. Somatic recombination consists of reciprocal, mitotic crossover, and nonreciprocal mitotic gene conversion assays. These tests measure DNA exchanges between segments of homologous chromatids. In essence, the processes produce homozygous states from heterozygous states in somatic cells. This process may be a factor in the expression of deleterious mutations for cancer or terata induction that are in a heterozygous state (+/−) in somatic cells.[34] However, the extrapolation of the yeast phenomena to mammalian chromosomes is rather tenuous.

### Insects

The somatic spot test using *D. melanogaster* described in this chapter can also be used to identify somatic cell recombination induced by genotoxic agents. Expression of recombination involves twin spots, which exhibit reciprocal events. These twin spots can be distinguished from single spots of altered wing hair morphology indicative of mutation or chromosome deletions.

### Cultured Mammalian Cells

Several types of tests for primary DNA damage have been developed using mammalian cells. The tests can be categorized into four basic meth-

TABLE 8.8
General Approaches to the Detection of DNA Damage Using Repair Endpoints

| Description of test | Typical cell type employed | Method for primary DNA damage detection |
|---|---|---|
| UDS | WI-38 cells or primary rat hepatocytes | Autoradiographic analysis of [$^3$H]thymidine incorporation into non-S-phase nuclei of treated cells |
| Alkaline elution technique | Any cell type | Detection of a reduction in DNA molecular size by elution of treated cells with alkaline on filters permitting small DNA species to pass through |
| DNA strand breakage | Any cell type | Detection of broken DNA strands by changes in DNA sedimentation in $CsCl_2$ gradients following treatment of cells with the test agent |
| $^{32}$P-Postlabeling assay | Any cell type | Detection of bulky DNA adducts by digesting DNA from exposed cells and labeling the mononucleotides with $^{32}$P; abnormal (adduct-containing) bases are observed as new spots on two-dimensional or four-dimensional TLC plates subjected to autoradiography (see Chapter 6 for details) |

ods (Table 8.8). The second and third tests in Table 8.8 are similar in that each measures the ability of the test chemical to break the DNA molecule in the chromosome, producing numerous smaller molecular weight pieces distinguishable from the normal high-molecular weight intact molecule. This is not analogous to chromosome breakage, since the effect is not cytologically visible and the damage is likely to be repaired. Neither of these tests is widely employed in routine screening; they are time-consuming and relatively susceptible to misinterpretation unless performed under stringent conditions by highly experienced investigators.

The DNA repair assay used by most genetic toxicology laboratories is the UDS technique. If the target cells are prevented from replicating (scheduled DNA synthesis) by nutrient deprivation or addition of hydroxyurea and are simultaneously exposed to the test material and $^3$H-labeled thymidine, no significant incorporation of labeled tritium will occur unless the test substance damages DNA, which stimulates the repair system (UDS). The stimulation of repair and subsequent incorporation of $^3$H-labeled thymidine can be detected via DNA extraction and scintillation analysis or by autoradiographic techniques (Figure 8.5). Due to

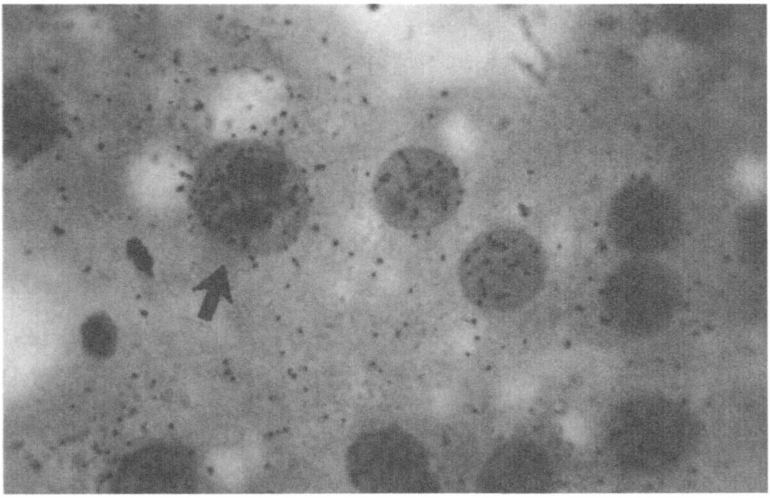

FIGURE 8.5. Exposed grains in photographic emulsion overlaying rat hepatocytes. Nuclei with dense grain counts indicate UDS.

the high degree of background noise of the scintillation technique, it is not recommended for general use.

An adaptation of the UDS autoradiographic methodology to rat primary hepatocytes has been developed by Williams.[35] This method offers two distinct advantages over the UDS assay employing WI-38 cells. In the technique developed by Williams, there is no need to use nutrient deprivation or chemical blockage of scheduled DNA synthesis, since the hepatocytes are nonreplicating over at least 48 hr after collection. In addition, the primary hepatocytes carry out most microsomal enzyme reactions, ensuring reliable detection of metabolic reaction products. This approach appears, initially, at least, to be the most reliable of the primary DNA damage assays in mammalian cells.

In 1980 an *in vivo* version of this test was described by Mirsalis and Butterworth.[36] This modification involves exposure of the rats *in vivo* followed by removal of hepatocytes from rats killed at various intervals postexposure. UDS analysis is conducted *in vitro* on the collected hepatocytes. Chemicals that require extrahepatic metabolic activation might be detected in this method, but not in the completely *in vitro* technique of Williams. The *in vivo*/*in vitro* technique has recently been reviewed by Ashby.[37]

## SCE Analysis

SCE analysis is a rapid, objective method of observing reciprocal exchanges between sister chromatids. Although the concept of SCE was established by Taylor in 1958,[38] recent staining techniques have greatly facilitated performance of this assay. The general method relies upon the phenomenon of bromodeoxyuridine (BrdU) incorporation into DNA in place of thymidine. After two rounds of cell division, the two chromatids are asymmetrically labeled with BrdU and consequently stain differently with the Hoeschst stain. SCEs are visualized as light–dark exchanges between chromatids stained after two mitotic divisions. While not leading to genetic toxicity *per se,* SCEs appear to respond to genotoxic compounds in a dose-related fashion, and can offer clear evidence of DNA exposure. The greatest use for SCE analysis is as a rapid, sensitive *in vitro/in vivo* method to show compound–chromosome interaction (dosimetry). Confirmatory data from other assays are usually necessary to support health concerns for SCE-inducers.

## *In Vivo* Mammalian Primary DNA Damage

Techniques to detect DNA damage *in vivo* are limited. The alkaline elution technique was applied by Swenberg and Petzold[39] to DNA carefully extracted from cells of various organisms in mice exposed to mutagenic agents. The methodology is amenable to this type of *in vivo* analysis, but test conditions must be very carefully controlled, since cellular damage (e.g., liposome disruption) resulting from treatment conditions or DNA collecting methods might introduce single- or double-strand breaks. Dose selection must be carefully controlled as well, since cytotoxicity will result in DNA breakage. These effects are independent of more relevant DNA-directed damage, and may lead to inappropriate assumption of genetic activity. This technique is technically difficult and will probably not be useful for chemical screening programs.

SCEs can be detected *in vivo* as well as *in vitro*. The critical aspect of *in vivo* studies is the necessity of having the target cells undergo two rounds of DNA replication in the presence of BrdU. Application of BrdU *in vivo* has been accomplished by intravenous infusion via a pump system or by subcutaneous implant of BrdU tablets.[40] The latter method is the simplest and requires only minor surgery. Once the tablet is implanted, the BrdU diffuses to many replicating sites within the animal. The animal is killed at a specified time, and the target cells removed and stained. If sufficient BrdU reached the target cells, visualization and scoring of SCE induction is possible. Highly proliferative cells, such as bone marrow, appear to be a good source of target cells.

Additional methods have been developed to collect cells from exposed animals and perform the staining techniques by culturing the cells in medium containing BrdU. One is a method developed by Stetka[41] in which peripheral lymphocytes are collected from the ear vein of a rabbit treated with the test agent. This method has the distinct advantage of permitting sequential samples from the same animal, permitting each animal to serve as its own control.

*In vivo* SCE evaluation in experimental animals is becoming a routine assay, and baseline data have been generated using the technique. Good baseline data for humans are also available. SCE appears to be a useful probe for *in vivo* analysis because of its sensitivity and scoring precision and will be used increasingly in dosimetry studies (see Chapter 6). While limited in application due to the lack of associated toxicity manifestations, SCE analysis offers an excellent opportunity to make direct response comparisons between animals and humans.

## *In Vitro* Cell Transformation

Cellular transformation *in vitro* appears operationally to simulate events that occur during the oncogenic transformation process *in vivo*. It is also probably the most controversial *in vitro* test method used in safety assessment. Transformation tests have many advantages as carcinogen screens, but have numerous disadvantages as well.[42] It appears that different transformation tests may be mechanistically dissimilar. The BHK-21 test originally proposed as a screen by Styles may in fact have measured mutation and not true cell transformation.[43] BALB/c–3T3 cells and C3H10T½ cells might be partially transformed and thus not give relevant information about test agents. Other systems, which are viral-dependent or viral-enhanced, probably involve complex relationships between chemicals and viral genes and will probably not be useful for general screening.

TABLE 8.9
Favorable Characteristics of BALB/c–3T3 and C3H10T½ Cell Transformation Assays

| |
|---|
| Endogenous metabolic capacity is reasonably good |
| Short expression time for transformed phenotype |
| Reasonably reproducible spontaneous transformation background |
| Target cells readily available |
| Amenable to the use of exogenous metabolic activation |
| Cells from foci can be transplanted into BALB/c or C3H mice for verification of transformation |
| Initiation and promotion models are available |

TABLE 8.10
Summary of *in Vitro* Cell Transformation Assays

| Target cell | Source | Key reference(s) | Features | Possible limitations |
|---|---|---|---|---|
| SHE | SHE fibroblast cells obtained from pooled litters, which are trypsinized to obtain cell suspensions | 24, 45, 46 | Primary or low-passage cells used as targets<br>Spontaneous background virtually zero[a]<br>Can be conducted as clonal or focus assay<br>Assay can be conducted in about 2 weeks<br>Nontransformed target cells nontumorigenic<br>Most type III foci demonstrate malignancy when transplanted to animals | Target cells must be preselected and tested; availability of target cells not predictable<br>Levels of induced transformation sometimes very low, making interpretation difficult<br>Clonal assay evaluation requires considerable experience, and scoring is often highly subjective |
| BALB/c–3T3 | BALB/c mouse fibroblast cells | 47, 48 | Spontaneous background relatively low or absent (C3H10T½) | Target cells are from a continous line, and therefore not equivalent to |

| | | | | |
|---|---|---|---|---|
| $C_3H10T½$ | $C_3H$ mouse fibroblast cells | 49, 50 | Induced numbers of foci generally high<br>Conducted as focus assays<br>Assays require about 4–6 weeks<br>Nontransformed target cell nontumorigenic<br>Most type III foci demonstrate malignancy when transplanted to appropriate animals<br>Amenable to exogenous activation samples such as S9<br>Cells are from continuous line and therefore readily available and should be similar from lab to lab<br>Scoring foci is easier than SHE | "normal" primary cells<br>Spontaneous background susceptible to change unless carefully controlled |
| Baby hamster BHK-21 cells | Baby hamster kidney fibroblast cells | 51 | Test can be conducted in 2 weeks<br>Transformation based on growth in soft agar and not morphology<br>Target cells are from a continuous line and readily available | Nontransformed cells have low level of tumorigenic potential<br>Spontaneous background generally very high<br>Growth in soft agar is quite variable, making test very difficult to score<br>Mechanism appears to be a single step; process like mutation rather than transformation |

[a] Recent evidence suggests that a low spontaneous transformation frequency does exist for these cells.

All of these issues have tended to prevent the development of an accepted screening test for cell transformation.

During the past few years, however, two transformation assays, the BALB/c–3T3 cell system and the C3H10T½ cell system, appear to have emerged as good choices for routine screening. There are several reasons for their emergence, some of which are given in Table 8.9. The data base for the BALB/c–3T3 assays is expanding rapidly and this test is probably the test of choice at this time.

### Initiation Assays

Ideally, human cells should be used as targets for transformation. Recent studies indicate that human cells can be transformed by chemicals, but the methodology needed for these assays is hardly ready to be applied in routine screening.[44] Only a few systems can be viewed as screening tests; they range from primary or passage cells to well-characterized cell lines. The most widely used assays are summarized in Table 8.10.

The addition of a metabolizing system to the transformation assays significantly enhances their utility in chemical screening. Recently, coupling target cells with S9 mix or primary hepatocyte-metabolizing cells has been successfully accomplished in the SHE, BALB/c–3T3, and 10T½ assays. This development has substantially expanded the utility of transformation assays by permitting detection of compounds with strict requirements for activation.[52–54]

### Promotion Assays

The mechanism of oncogenicity induced by a chemical may be important to how exposures to such agents are regulated. Since it could be argued that agents that initiate tumors might have different induction kinetics that agents that are pure promoters, it is of extreme value to be able to recognize promoters among the classes of chemicals tested.[55] The following is a list of agents suspected of having promotion, but no initiation capability:

| | |
|---|---|
| Phorbol esters (TPA and derivatives) | DDT |
| Phenobarbital | Alkanes |
| Butylated hydroxytoluene (BHT) | Bile salts |
| Anthralin | |

The typical BALB/c–3T3 transformation assay consists in an initial exposure of 24–72 hr followed by removal of the medium containing the test sample and biweekly media changes. At the end of 4–6 weeks, the cells have become confluent, and if the initial exposure was transforming,

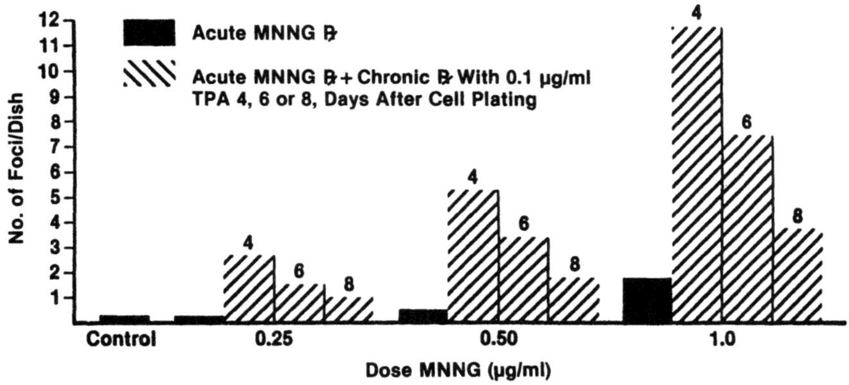


FIGURE 8.6. Promoting activity of TPA on MNNG-initiated BALB/c–c3T3 cells. The results show that the promoting function is enhanced as the initiating dose is increased; the optimal time for beginning TPA treatment is 4 days after cell seeding. (J. O. Rundell and E. Matthews, unpublished data.)

foci of morphologically distinct cells are visible. A promoting study consists in repeated exposures of the suspect promoter subsequent to a subtransforming dose of a positive carcinogen. Results using a model initiator and promoter are given in Figure 8.6. The addition of low, nontransforming levels of TPA (0.1 μg/ml) to MNNG-initiated cells results in a greatly enhanced level of transformed foci. In addition, the data show two other characteristics similar to *in vivo* two-stage studies, in that the initiation process is dose-dependent at a constant level of promoter and the degree of enhancement diminishes as the time between initiation and application of the promoter increases. Additional studies with this model have demonstrated that TPA analogs that do not promote *in vivo* are nonpromoting in the *in vitro* model and that an optimum time to apply the promoter following initiation exists (Table 8.11).

Another *in vitro* model to detect promoters has been proposed, which depends upon chemical-mediated alterations in cell-to-cell contact as a technique to identify potential promoting agents. This assay system is based on observations of reduced recovery of 6-thioguanine-resistant ($HGPRT^-$) mutant cells from a mixed population of normal 6-thioguanine-sensitive ($HGPRT^+$) and mutant cells due to metabolic cooperation. The cell-to-cell contact between the normal cell and the mutant cell permits transport of a toxic intermediate of 6-thioguanine into the genetically resistant mutant cell.

Transport between cells occurs only under conditions favoring the formation of gap junctions. Trosko and his colleagues have found that known promoting agents inhibit gap-junction formation and thus enhance the recovery of HGPRT mutant cells from a mixed mutant/nonmutant

TABLE 8.11
The Results of a Comparison between *in Vivo* (Skin Painting) Promotion and *in Vitro* Promotion for Phorbol and Phorbol Diesters

| Test material | Reported *in vivo* promotion activity | Activity in *in vitro* promotion assay[a] |
|---|---|---|
| Phorbol-12,13-didecanoate | + | + |
| 12-*O*-Tetradecanoyl-phorbol-13-acetate (TPA) | + | + |
| 4-*O*-Methyl-12-*O*-tetradecanoyl-phorbol-13-acetate | − | − |
| Phorbol | − | − |
| 4-Phorbol-12,13-didecanoate | − | − |

[a] This procedure was developed for the BALB/c–3T3 assay and basically consists of a chronic treatment regimen in the potential promoting agent. The promotor TPA is used as a control in the study design. (Results courtesy of Dr. E. Matthews.)

population exposed to 6-thioguanine[56] (Figure 8.7). While this technique does not mimic the *in vivo* study design, it does fit with proposed models of how promoting agents function. Cell-to-cell contact might well be important in controlling the transition of initiated cells to the transformed condition, and agents that interfere with normal cellular communication might act to release initiated cells into a transformed mode.

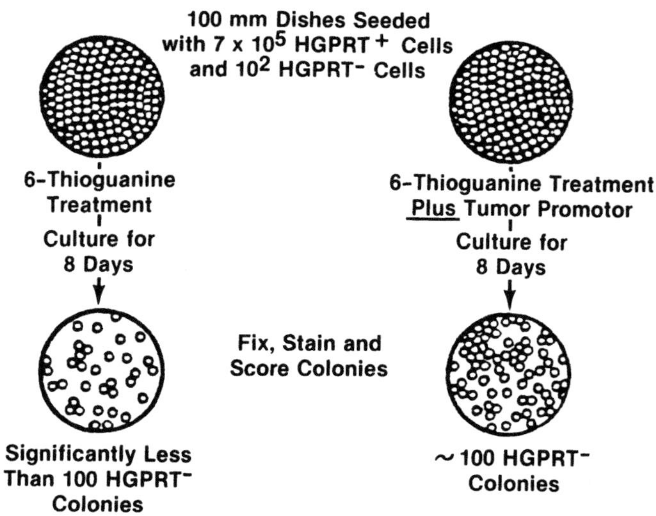


FIGURE 8.7. Metabolic cooperation assay for suspect promoting agents.

## Transfection Assays

This assay evaluates the morphologic transforming potential of biologicals, using cultured mammalian cells as targets. While the transformation assay has been commonly used to test chemicals for transforming activity, it may also be used to detect the transforming potential of samples of DNA in what are commonly referred to as "DNA transfection experiments." Both NIH–3T3 and BALB/c–3T3 cell lines have been demonstrated to be transformed by DNA samples from appropriate mammalian cell sources.

The DNA transfection procedure involves a calcium phosphate precipitation technique first described by Graham and van der Eb.[57] It has been successfully used to demonstrate that MCA-transformed BALB/c–3T3 cells contain c-ki-*ras* oncogenes. DNA extracted from this transformed cell line was shown to induce a morphologic transformation of NIH–3T3 cells.[58,59] Similarly, Hopkins *et al.*[60] have used this procedure to perform DNA transfection experiments in transformable BALB/c–3T3 cells.

In the DNA-transfection transformation assay, the test material is commonly a high-molecular-weight DNA specimen. In most protocols, the BALB/c–3T3 cells will be pretreated with the DNA transfection buffer solution and DNA specimens for a total of ~5 hr.[57,58] This 5-hr treatment includes an adsorption period of ~20 min and additional 4-hr incubation period. Following the treatment, the test material and positive control DNA samples are removed and the cultures are refed with culture medium. The cell cultures are refed twice weekly for an additional 3–4 weeks to allow for expression of the transforming activity. Assays are terminated and evaluated using the same procedures for foci visualization and criteria that have been developed and applied to the standard BALB/c–3T3 cell transformation assay.

## REFERENCES

1. Reddy, E. P., Reynolds, R. K., Santos, E., and Barbacid, M.: A point mutation is responsible for the acquisition of transforming properties by the T24 human bladder carcinoma oncogene. *Nature* **300**:149–150, 1982.
2. Brown, M. M., Wassom, J. S., Malling, H. V., Shelby, M. D., and Von Halle, E. S.: Literature survey of bacterial, fungal, and *Drosophila* assay systems used in the evaluation of selected chemical compounds for mutagenic activity. *J. Natl. Cancer Inst.* **62**(4):841–871, 1979.
3. Brusick, D.: Bacterial mutagenesis and its role in the identification of potential animal carcinogens. In *Carcinogens: Identification and Mechanisms of Action* (A. C. Griffin and C. R. Shaw, eds.), Raven Press, New York, pp. 93–105, 1979.
4. Ames, B. N., McCann, J., and Yamasaki, E.: Methods for detecting carcinogens and

mutagens with the *Salmonella*/mammalian-microsome mutagenicity test. *Mutat. Res.* **31**:347–364, 1975.

5. Maron, D., and Ames, B. N.: Revised methods for the *Salmonella* mutagenicity test. *Mutat. Res.* **113**:173–215, 1983.
6. Levin, D. E., Hollstein, M., Christman, M. F., Schwiers, E. A., and Ames, B. N.: A new *Salmonella* tester strain (TA102) with A–T base pairs at the site of mutation detects oxidative mutagens. *Proc. Natl. Acad. Sci. USA* **79**:7445–7449, 1982.
7. Brusick, D. J.: Observations and recommendations regarding routine use of bacterial mutagenesis assays as indicators of potential chemical carcinogens. In *Strategies for Short-Term Testing for Mutagens/Carcinogens* (B. E. Butterworth, ed.), CRC Press, West Palm Beach, Florida, pp. 3–12, 1979.
8. Hollstein, M., McCann, J., Angelosanto, F., and Nichols, W.: Short-term tests for carcinogens and mutagens. *Mutat. Res.* **65**:133–226, 1979.
9. Brusick, D. J.: Induction of cycloheximide-resistant mutants in *Saccharomyces cerevisiae* with *N*-methyl-*N*′-nitro-*N*-nitrosoguanidine and ICR-170. *J. Bacteriol* **109**(3):1134–1138, 1972.
10. Callen, D. F., and Philpot, R. M.: Cytochrome P-450 and the activation of promutagens in *Saccharomyces cerevisiae*. *Mutat. Res.* **45**:309–324, 1977.
11. Tindall, K. R., Stankowski, Jr., L. F., Machanoff, R., and Hsie, A. W.: Detection of deletion mutations in pSV2*gpt*-transformed cells. *Mol. Cell. Biol.* **4**(7):1411–1415, 1984.
12. Vogel, E.: Identification of carcinogens by mutagen testing in *Drosophila:* The relative reliability for the kinds of genetic damage measured. In *Origins of Human Cancer,* Vol. C, Cold Spring Harbor Conferences on Cell Proliferation, Cold Spring Harbor Laboratory, Cold Spring Harbor, New York, pp. 1483–1497, 1977.
13. Vogel, E., and Sobels, F. H.: The function of *Drosophila* in genetic toxicology testing. In *Chemical Mutagens: Principles and Methods for Their Detection,* Vol. 4 (A. Hollaender, eds.), Plenum Press, New York, pp. 93–142, 1976.
14. Würgler, F. E., Sobels, F. H., and Vogel, E.: *Drosophila* as assay system for detecting genetic changes. In *Handbook of Mutagenicity Test Procedures* (B. J. Kilbey, M. Legator, W. Nichols, and C. Ramel, eds.), Elsevier/North-Holland, Amsterdam, pp. 335–373, 1977.
15. Hällstrom, I., Magnusson, J., and Ramel, C.: Relation between the somatic toxicity of dimethylnitrosamine and a genetically determined variation in the level and induction of cytochrome P450 in *Drosophila melanogaster, Mutat. Res.* **92**:161–168, 1982.
16. Boyd, J. B., and Setlow, R. B.: Characterization of post-replication repair in mutagen-sensitive strains of *Drosophila melanogaster*. *Genetics* **84**:507–526, 1976.
17. Graf, U., Juon, H., Katz, A. J., Frei, H. J., and Würgler, F. E.: A pilot study on a new *Drosophila* spot test. *Mutat. Res.* **120**:233–239, 1983.
18. Russell, L. B., Selby, P. B., van Halle, E., Sheridan, W., and Valcovic, L.: The mouse specific locus test with agents other than radiations: Interpretation of data and recommendations for future work. *Mutat. Res.* **86**:329–354, 1981.
19. Ehling, U. H.: Specific-locus mutations in mice. In *Chemical Mutagens: Principles and Methods for Their Detection,* Vol. 5 (A. Hollaender and F. J. de Serres, eds.), Plenum Press, New York, pp. 233–256, 1978.
20. Fahrig, R.: The mammalian spot test: A sensitive *in vivo* method for the detection of genetic alterations in somatic cells of mice. In *Chemical Mutagens: Principles and Methods for Their Detection,* Vol. 5 (A. Hollaender and F. J. de Serres, eds.), Plenum Press, New York, pp. 151–176, 1978.
21. Kratochvilova, J., and Ehling, U. H.: Dominant cataract mutations induced by γ-irradiation of male mice. *Mutat. Res.* **63**:221–223, 1979.
22. Ehling, U. H.: Risk estimations based on germ-cell mutations in mice. In *Environmental*

*Mutagens and Carcinogens (Proceedings of the 3rd International Conference on Environmental Mutagens)* E. T. Sugimura, S. Kondo, and H. Takebe, eds.), University of Tokyo Press, Tokyo and Alan R. Liss, New York, pp. 709–719, 1982.
23. Selby, P. B., and Lee, S. S.: Sensitive-indicator results show that ethylnitrosurea is also a super mutagen for dominant skeletal mutations. *Environ. Mutag.* **3**:373, 1981.
24. Styles, J. A., and Penman, M. P.: The mouse spot test: An appraisal of its performance in detecting chemical carcinogens and mutagens. *Mutat. Res.* **154**:183–204, 1985.
25. Bignami, M., Morpurgo, G., Agliani, R., Carere, A., Conti, G., and Di Giuseppe, G.: Non-disjunction and crossing-over induced by pharmaceutical drugs in *Aspergillus nidulans. Mutat. Res.* **26**:159–170, 1974.
26. Parry, J. M., and Zimmermann, F. K.: The detection of monosomic colonies produced by mitotic chromosome non-disjunction in the yeast *Saccharomyces cerevisiae. Mutat. Res.* **36**:49–66, 1976.
27. Ehling, U. H., Machemer, L., Buselmaier, W., Dýcka, J., Frohberg, H., Kratochvilova, J., Lang, R., Lorke, D., Müller, D., Peh, J., Rohrborn, G., Roll, R., Schulze-Schencking, M., and Wiemann, H.: Standard protocol for the dominant lethal test on male mice. Set up by the work group, Dominant Lethal Mutations of the *Ad Hoc* Committee on Chemogenetics. *Arch. Toxicol.* **39**:173–185, 1978.
28. Holden, H.: Comparison of somatic and germ cell models for cytogenetic screening. *J. Appl. Toxicol.* **2**:196–200, 1982.
29. Generoso, W. M., Cain, K. T., Huff, S. W., and Gosslee, D. G.: Heritable translocation test in mice. In *Chemical Mutagens: Principles and Methods for Their Detection,* Vol. 5 (A. Hollaender and F. J. de Serres, eds.), Plenum Press, New York, 1978.
30. Wiemann, H., and Lang, R.: Strategies for detecting heritable translocations in male mice by fertility testing. *Mutat. Res.* **53**:317–326, 1978.
31. Carrano, A. V., Thompson, L. H., Lindl, P. A., and Minkler, J. L.: Sister chromatid exchange as an indicator of mutagenesis. *Nature* **271**:551–553, 1978.
32. Rosenkranz, H. S., and Leifer, Z.: Determining the DNA-modifying activity of chemicals using DNA polymerase-deficient *Escherichia coli.* In *Chemical Mutagens: Principles and Methods for Their Detection,* Vol. 6 (A. Hollaender and F. J. de Serres, eds.), pp. 109–147, Plenum Press, New York, 1980.
33. Kada, T., Hirano, K. and Shirasu, Y.: Screening of environmental chemical mutagens by the Rec-assay system with *Bacillus subtilis.* In *Chemical Mutagens: Principles and Methods for Their Detection,* Vol. 6 (A. Hollaender and F. J. de Serres, eds.), pp. 149–173, Plenum Press, New York, 1979.
34. Zimmermann, F. K.: Procedures used in the induction of mitotic recombination and mutation in the yeast *Saccharomyces cerevisiae. Mutat. Res.* **31**:71–86, 1975.
35. Williams, G. M.: The detection of chemical mutagens/carcinogens by DNA repair and mutagenesis in liver cultures. In *Chemical Mutagens: Principles and Methods for Their Detection,* Vol. 6 (A. Hollaender and F. J. de Serres, eds.), Plenum Press, New York, pp. 61–79, 1980.
36. Mirsalis, J. C., and Butterworth, B. E.: Detection of unscheduled DNA synthesis in hepatocytes isolated from rats treated with genotoxic agents: An *in vivo–in vitro* assay for potential carcinogens and mutagens. *Carcinogenesis* **1**:621–625, 1980.
37. Ashby, J., Lefevre, P. A., Burlinson, B., and Penman, M. G.: An assessment of the *in vivo* rat hepatocyte DNA-repair assay. *Mutat. Res.* **156**:1–18, 1985.
38. Taylor, J. H.: Sister chromatid exchange in tritium-labeled chromosomes. *Genetics* **43**:515–529, 1958.
39. Swenberg, J. A., and Petzold, G. L.: The usefulness of DNA damage and repair assays for predicting carcinogenic potential of chemicals. In *Strategies for Short-Term Testing for Mutagens/Carcinogens* (B. E. Butterworth, ed.), CRC Press, West Palm Beach, Florida, pp. 77–86, 1979.

40. Allen, J. W., Shuler, C. F., Mendes, R. W., and Latt, S. A.: A simplified technique for *in vivo* analysis of sister chromatid exchanges using 5-bromodeoxyuridine tablets. *Cytogenet. Cell Genet.* **18**:231–237, 1977.
41. Stetka, D. G.: Sister chromatid exchange in animals exposed to mutagenic carcinogens—An assay for genetic damage in humans. In *Proceedings of the 9th Annual Conference on Environmental Toxicology,* Aerospace Medical Research Laboratory, Dayton, Ohio, pp. 215–222, 1979.
42. Rundell, J. O.: *In Vitro* cell transformation: An overview. In *Carcinogenesis and Mutagenesis Testing,* J. F. Douglas, ed., Humana Press, Clifton, New Jersey, pp. 39–62, 1984.
43. Styles, J. A.: Mammalian cell transformation *in vitro*. Six tests for carcinogenicity. *Br. J. Cancer* **37**(6):931–936, 1978.
44. Kakunaga, T.: The transformation of human diploid cells by chemical carcinogens. In *Origins of Human Cancer,* Vol. 4 (H. H. Hiatt, J. D. Watson, and J. A. Winsten, eds.), Cold Spring Harbor Conferences on Cell Proliferation, Cold Spring Harbor Laboratory, Cold Spring Harbor, New York, pp. 1537–1548, 1977.
45. DiPaolo, J. A., Donovan, P., and Nelson, R. L.: Quantitative studies of *in vitro* transformation by chemical carcinogens. *J. Natl. Cancer Inst.* **42**:867–874, 1969.
46. DiPaolo, J. A., Nelson, R. L., and Donovan, P. J.: Morphological, oncogenic, and karyological characteristics of Syrian hamster embryo cells transformed *in vitro* by carcinogenic polycyclic hydrocarbons. *Cancer Res.* **31**:1118–1127, 1971.
47. DiPaolo, J. A., Takano, K., and Popescu, N. C.: Quantitation of chemically induced neoplastic transformation of BALB/3T3 cloned cell lines. *Cancer Res.* **32**:2686–2695, 1972.
48. Kakunaga, T.: Quantitative system for assay of malignant transformation by chemical carcinogens using a clone derived from BALB/3T3. *Int. J. Cancer* **12**:463–473, 1973.
49. Chen, T. T., and Heidelberger, C.: Quantitative studies on the malignant transformation *in vitro* of cells derived from adult C3H mouse ventral prostate. *Int. J. Cancer* **4**:166–167, 1969.
50. Marquardt, H., Juroki, T., Huberman, E., Selkirk, J. K., Heidelberger, C., Grover, P. L., and Sims, P.: Malignant transformation of cells derived from mouse prostate by epoxides and other derivatives of polycyclic hydrocarbons. *Cancer Res.* **32**:716–720, 1972.
51. Purchase, I. F. H., Longstaff, E. A., Ashby, J., Styles, J. A., Anderson, D., Lefevre, P. A., and Westwood, F. R.: Evaluation of six short term tests for detecting organic chemical carcinogens and recommendations for their use. *Nature* **264**:624–627, 1976.
52. Curren, R. D., Kouri, E., and Schechtman, L. M.: Studies on metabolic activation of chemicals for mammalian cell transformation and mutagenesis. In *Advances in Modern Environmental Toxicology, Vol. I, Mammalian Cell Transformation by Chemical Carcinogens* (N. Mishra, V. Dunkel, and M. Mehlman, eds.), Chapter 12, pp 319–353, 1981.
53. McGlynn-Kreft, A. Scriner, H. E., and McCarthy, K. L.: An exogenous metabolic activation system for the C3H10T½ cell transformation assay. 13th Annual Meeting of the Environmental Mutagen Society, Abstract B3-3, 1983.
54. Pienta, R. J.: A hamster embryo cell model system for identifying carcinogens. In *Carcinogens: Identification and Mechanisms of Action* (A. C. Griffin and C. R. Shaw, eds.), Raven Press, New York, pp. 121–141, 1979.
55. Sivak, A.: An evaluation of assay procedures for detection of tumor promotors. *Mutat. Res.* **98**:377–387, 1982.
56. Trosko, J. E., Dawson, B., Yotti, L. P., and Change, C. C.: Saccharin may act as a tumor promotor by inhibiting metabolic cooperation between cells. *Nature* **285**:109–110, 1980.

57. Graham, F. L., and van der Eb, A. J.: A new technique for the assay of infectivity of human adenovirus 5 DNA. *Virology* **52**(2):456–467, 1973.
58. Shih, C., Shilo, B. Z., Goldfarb, M. P., Dannenberg, A., and Weinberg, R. A.: Passage of phenotypes of chemically transformed cells via transfection of DNA and chromatin. *Proc. Natl. Acad. Sci. USA* **76**:5714–5718, 1979.
59. Shilo, B. Z., and Weinberg, R. A.: Unique transformation gene in carcinogen-transformed mouse cells. *Nature* **289**:697–709, 1981.
60. Hopkins, N., Besmer, P., Deleo, A. B., and Law, L. W.: High-frequency cotransfer of the transformed phenotype and a tumor specific transplantation antigen by DNA from the 3-methycholanthene-induced Meth A sarcoma of BALB/c mice. *Proc. Natl. Acad. Sci. USA* **78**:7555–7559, 1981.

CHAPTER 9

# Applications from Biotechnology Research to Genetic Toxicology

## INTRODUCTION

Technological breakthroughs in nucleic acid synthesis, gene cloning restriction enzyme analysis, and monoclonal antibody production will lead to significant improvements in molecular toxicology methods for testing and population monitoring. These improvements reside primarily in the specificity and quantitation of detection of toxicologically relevant DNA alterations and in the improved ability to measure those alterations *in vivo*.

Biotechnology research has produced a totally new type of industry.[1] This new technology emphasizes the biological production of new drugs and chemicals rather than by chemical synthesis. Consequently, one might expect that (1) the current array of test methods used to assess the safety of products may need revisions to accommodate bioengineered materials, and (2) the nature of the ecological impact of waste products and fugitive emissions from this new technology may be quite different from that released by conventional industries. U. S. FDA and EPA regulators are reviewing policies and developing guidelines that address these biotechnology-related issues.[2,3]

Biotechnology research is also beginning to lead to changes in the methodology of measuring DNA damage and the array of assays employed to assess mutagenic safety. It is likely that during the next 10 years the types of tests used today will be totally replaced with improved methods.

## IMPACT ON TEST METHODS

### Techniques of Interest

Among the rapidly evolving methods of biotechnology, three techniques appear promising as components of molecular toxicology procedures. The three are the construction of DNA probes, gene cloning, and production of monoclonal antibodies.

### Production of DNA Probes and Synthetic Nucleotide Sequences

The ability to rapidly purify and sequence (amino acid) gene products followed by the automated synthesis of a nucleotide sequence linked to the gene product structure will permit the production of a library of gene sequences that can be labeled (radiolabeling or immunolabeling). These probes may then be used to locate the position of genes on a chromosome by nucleotide hybridization between complementary sequences or to identify the existence of mutant genes by combining the probe with Southern DNA blotting techniques.

Many methods are available to acquire specific DNA probes for hybridization analysis. The following two methods are probably most common:

1. The nucleotide sequence of the normal gene may be known as a result of previous sequencing analysis. Consequently, the probe DNA can be synthesized rapidly by an automated DNA synthesizer. The intended application of the probe will determine its length in number of nucleotides. At least 500 base pairs are needed in probes for good site recognition and binding as diagnostic reagents. On the other hand, mutant sequence recognition requires shorter probes (19 base pairs) coupled with modified DNA binding temperatures to screen for mutagenic damage.

2. If the cellular gene product is easily accessible, one would also expect to find large numbers of specific mRNA molecules. mRNA can be isolated and converted to a complementary DNA (cDNA) copy, using an enzyme called reverse transcriptase. This enzyme is obtained from RNA retroviruses (RNA tumor viruses) and is important for conversion of viral RNA to DNA before insertion into the host cell genome. The cDNA can be prepared rather easily *in vitro* and used as the DNA probe. This method of probe production does not require knowledge of specific nucleotide sequences. The specificity of cDNA probes may not be as high as probes made by direct synthesis.

The methods of choice for detecting hybridized probes are the autoradiographic detection of radiolabeled nucleotides incorporated into the probe DNA sequence and the use of fluorescent antibody techniques.

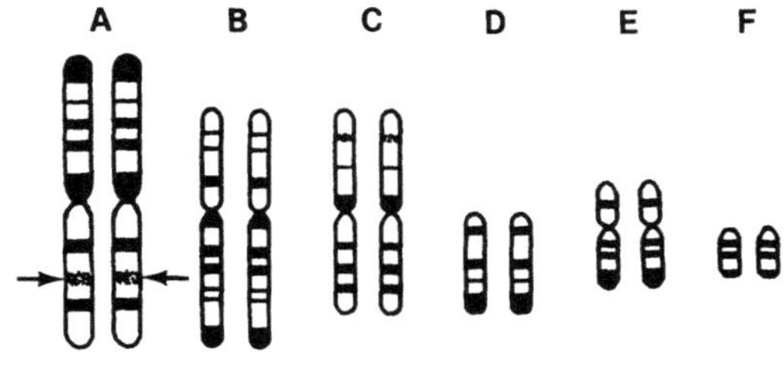


FIGURE 9.1. A radiolabeled DNA probe made from the cDNA of a specific gene was added to cells prior to karyotyping. Once the karyotype is developed, localization of the cDNA probe is possible, thus permitting localization of the primary site for the gene on one of the chromosomes (lower arm of chromosome A).

Radiolabeled sequences can be detected by analysis of the intact cells or of purified DNA from the cells by radioactive decay using photographic emulsions. The use of fluorescent antibody techniques has been made more attractive with the development of procedures to enhance the detection of fluorescent tags. During the past 3–5 years, antibody stacking and computer-assisted fluorescent enhancement have resulted in detection of fluorescent-labeled DNA probes at the single-cell level.[4] Detection of a single gene copy per cell is possible with these improvements. As a result, single cell populations obtained directly from individuals can be scored with specific DNA probes, for inherited mutant alleles or for induced sequence alterations.

## Applications of DNA Probes in Genetic Toxicology

DNA probes may be very useful in locating genes on one of the chromosomes of a karyotyped cell. The approach involves collection and purification of the mRNA from the gene to be mapped. From the mRNA, cDNA copies are made using a radiolabeled nucleotide. The cDNA is hybridized to the cellular DNA and standard banded karyotypes are prepared from cells subjected to autoradiographic techniques. The resultant karyotype will allow for gene localization to a banded area(s) via the cluster of exposed silver grains (Figure 9.1).

Shorter DNA probes of approximately 19 nucleotides can be used to detect mutations at single base pairs. The technique involves a combination hybridization and Southern blotting. Restriction enzymes characteristically cleave DNA at specific nucleotide sequences, resulting in a set of restriction fragments (RF). For a given organism treated with one or a combination of restriction enzymes the RF will always be identical. Thus, individual genes can be concentrated by treating DNA from cell pools with a restriction enzyme and separating the fragments by gel electrophoresis. The genes on a specific fragment will be concentrated in a

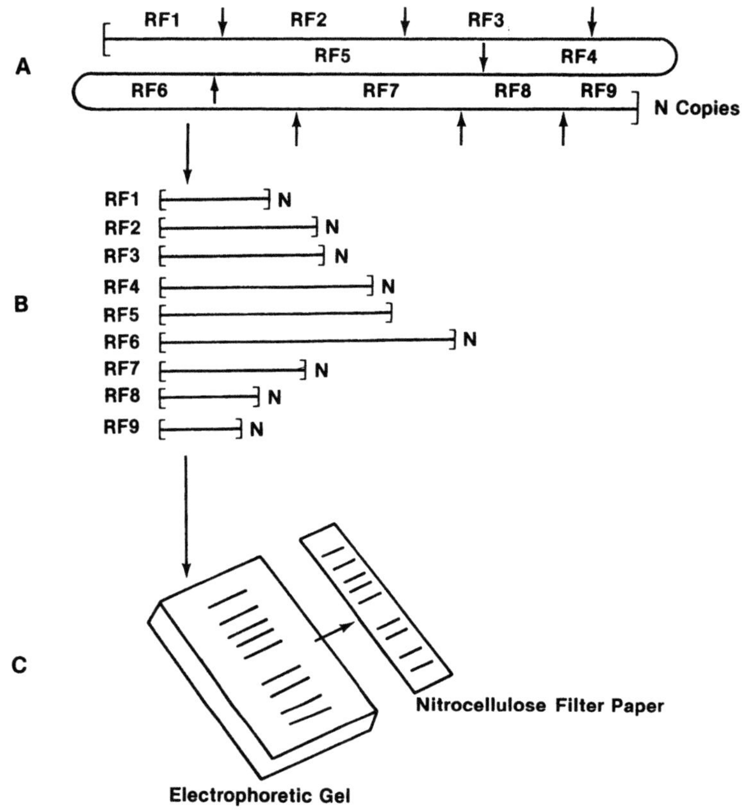


FIGURE 9.2. (A) Intact DNA is cleaved into restriction fragments (RF) by treatment with a mixture of restriction enzymes (B). Each set of RFs form a monospecific pool, which can be separated from other monospecific pools by gel electrophoresis (C). DNA RFs isolated and separated by gel electrophoresis are transferred from the gel to nitrocellulose filter paper strip. This step concentrates each monospecific pool of DNA into the paper strip, where it can then be hybridized to probe DNA.

monospecific band (Figure 9.2). In order to hybridize the DNA probe with the isolated DNA fragments, the DNA in the gels must be transferred to a more suitable environment for DNA–DNA interactions. This step consists in placing a nitrocellulose filter on top of the electrophoresis gel and passing a buffer solution through in such a way that the DNA is drawn from the gel to the nitrocellulose filter (Figure 9.2). Once deposited in the filter, the DNA fragments are then amenable for hybridization.

Hybridization conditions are critical, but usually are performed at about 40°C. The radiolabeled DNA probe will hybridize with the DNA band containing a complementary sequence (Figure 9.3). Mutation anal-

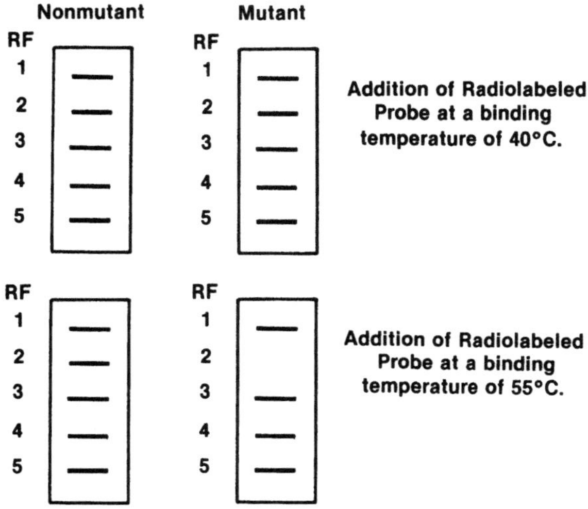


FIGURE 9.3. Southern blot strips from hybridization studies conducted at 40 and 55°C. Small (15–20 nucleotide) probes will hybridize to the mutant sequence at 40°C, but not at an abnormal hybridization temperature of 55°C. The probes are made to the nonmutant sequences and bind to nonmutant DNA at both temperatures. The slight deviation in the nucleotide sequence of the mutant reduces the hybridization efficiency at 55°C. This technique can be used with colonies or cell clones, but not with individual cells or mixed populations.

ysis involves modification of the hybridization conditions (temperature) such that the probe containing a sequence different from that found in the RF will not hybridize. Even if the difference is only a single base-pair substitution, hybridization will not occur at the restrictive temperature. In this fashion, DNA probes seeking specific mutations can be constructed. This technique is applicable to basic mutational analysis as well as fetal and newborn monitoring for specific mutant (e.g., hemoglobin) sequences.[5]

## Gene Cloning

This technique involves the collection, processing, insertion, and expression of specific genes or gene clusters (operons) in host organisms foreign to the origin of the expressed gene. The primary objective is to identify and recover useful genes with their associated regulatory sequences and amplify the production of product from that gene by placing it into an organism that can be efficiently cultured in high volumes at low cost. The result is mass production of a specific molecule that can be easily recovered.

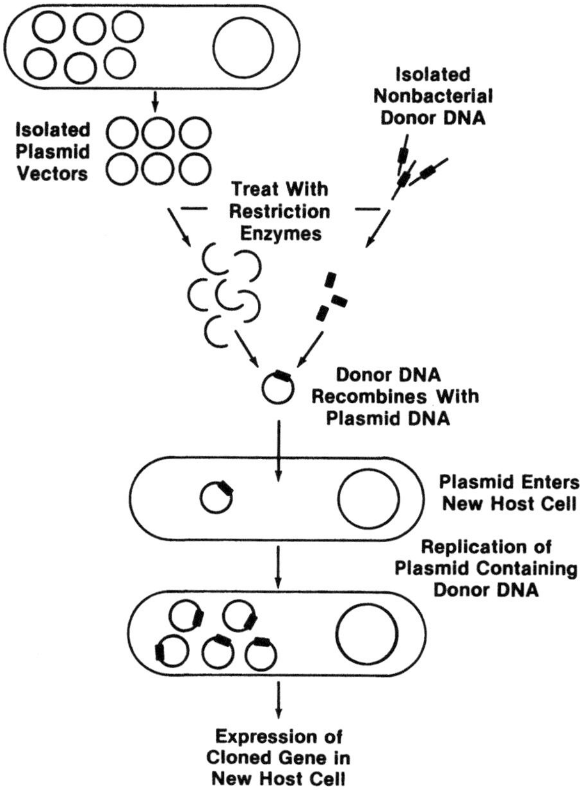


FIGURE 9.4. General steps of gene cloning in bacteria.

One of the most critical components in the gene cloning process is the availability of a suitable vector. A vector transports the cloned DNA, also identified as rDNA, into the new host.[6] In many cases, if not most, the vector is a natural component or symbiote of the host organism. For example, bacterial plasmids (circular double-standard, self-replicating pieces of DNA) are used as vectors for cloning rDNA in bacteria. Nonbacterial rDNA sequences can be inserted into the plasmid DNA as shown in Figure 9.4. Once inserted, plasmids will "infect" the host bacterium and coexist with the chromosomal DNA. The number of copies of plasmids varies, but by using a plasmid vector that is carried at a high multiplicity by the host bacteria, the level of gene product produced can be considerably enhanced.

Vectors can also include viruses that infect, but do not kill, the host cell. Incomplete viruses consisting of the infective and gene regulation

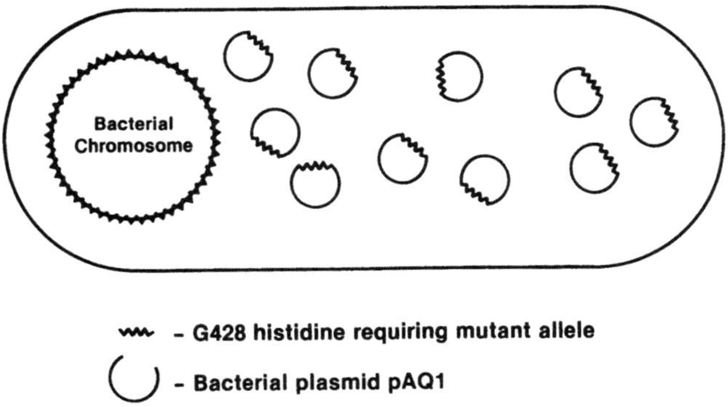


FIGURE 9.5. Application of gene cloning to the development of new *Salmonella typhimurium* tester strains. Representation of a bacterium of *S. typhimurium* strain TA102. The G428 *his* site is deleted from the bacterial chromosome and located on the bacterial plasmid. Reversion of the mutant site located on any one of the plasmids will result in a histidine-independent cell. This application of gene cloning results in a highly sensitive tester strain, since each cell contains many copies of the target DNA rather than the normal one copy per cell.

sequences are common vectors for carrying rDNA into mammalian cells or plant cells. Fungal and algal cells also have independently replicating DNA sequences, which can be used as vectors for cloning in these organisms. For commercial product development, bacteria and yeast are the most common host cells, although large-scale mammalian cell culture is possible using large fermentors and hollow-fiber, continuous-culture units.

## Application of Gene Cloning to Genetic Toxicology

An example of how this technology can be applied to genetic toxicology is seen in one of the new *Salmonella typhimurium* strains introduced by Maron and Ames.[7] Strain TA102 carries the *his* G428 mutation on the plasmid PAQ1. This plasmid exists at a high multiplicity, thus affording many copies of the target sequence for mutagens to interact with and revert. Reversion of only a single *his* mutation on one plasmid will yield a revertant colony (Figure 9.5). Other examples of this use of cloning include mutation assays with mammalian cell lines engineered to contain a microbial gene sequence that is easily recovered from the cell and analyzable for DNA alterations or that responded more sensitively to mutagens than the mammalian gene replaced by the microbial sequence.[8]

The most common approach is to acquire a gene from a eukaryotic

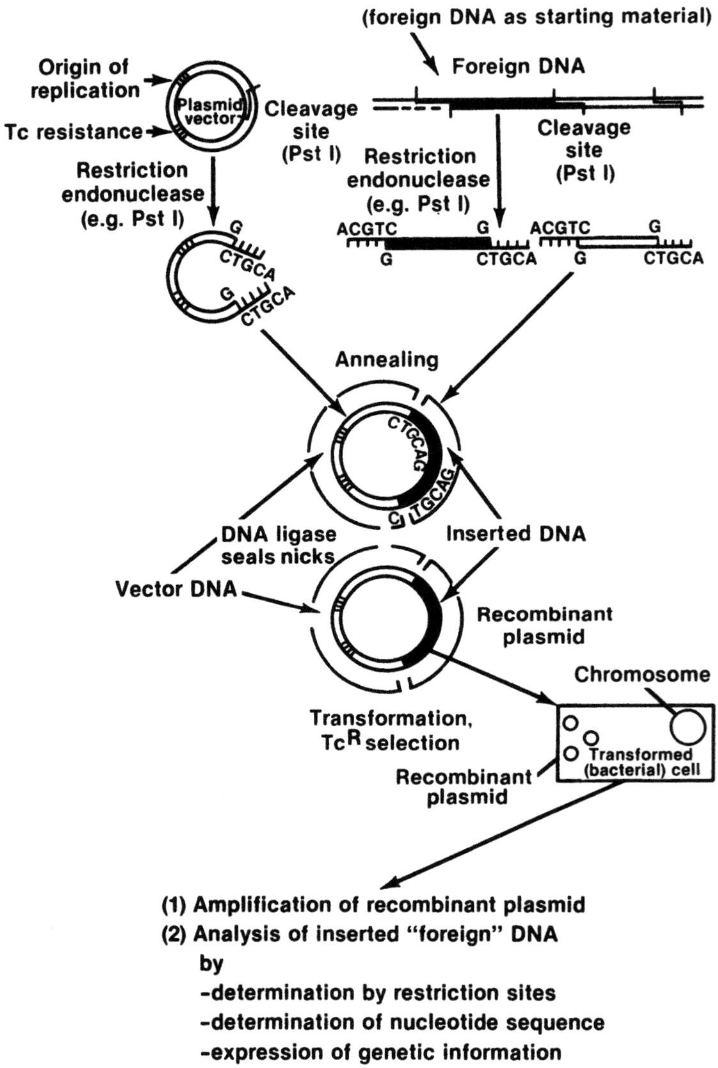


FIGURE 9.6. Biochemical/biological steps involved in nucleic acid cloning. (Courtesy of P. Lohman.)

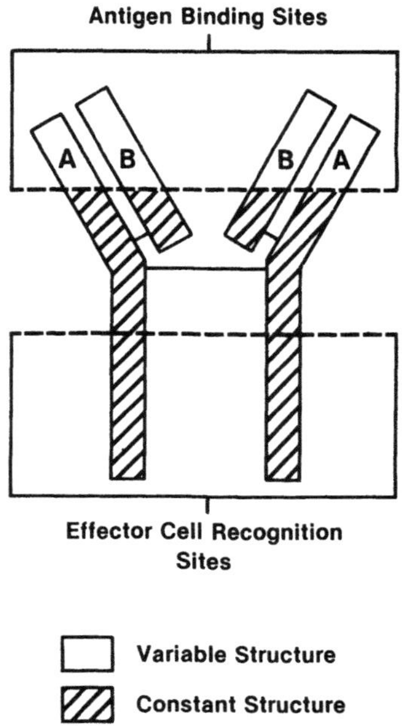


FIGURE 9.7. Antibody structure. The antibody molecule is composed of (A) heavy and (B) light chains. The variable structure interacts with an antigen and commits the antibody-binding site. The effector site is relatively constant and interacts with specific target cells of the host's immune system.

cell (generally slow-growing) and place the gene into a bacterial plasmid, which is then introduced into a host bacterium, where the plasmid is amplified and the host organism can be grown at high density in industrial fermentors. Figure 9.6 illustrates the detailed methods employed in inserting a mammalian gene into a bacterial plasmid. Cloning of genes in fungi (i.e., yeast), algae, or plants (crop plants) also has important implications in product development and new crop production.[9]

The production of DNA sequences for cloning is similar to the preparation of sequences for use as DNA probes. Synthetic gene sequences can be prepared via several methods or cDNA copies can be made from purified mRNA.

DNA sequences derived from these two methods may lack associated regulatory sequences necessary for regulation of gene expression. To overcome this difficulty, another technique that has been commonly used is the shotgun approach. In this approach, DNA libraries from a given organism are made by exposing the intact genome of the organism to a wide range of restriction endonuclease enzymes, which cut DNA at site-specific nucleotide sequences. The endonuclease cuts produce an array of DNA pieces of various lengths. The complete RF composite of the entire genome of an organism constitutes that organism's DNA library. Among the "library" of DNA sequences, one will exist that contains not

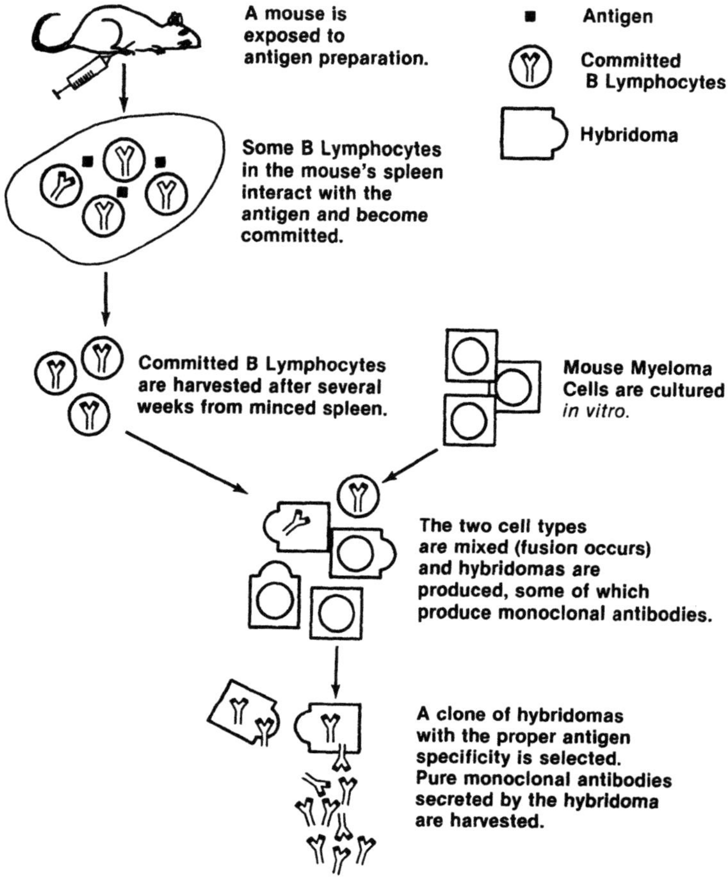


FIGURE 9.8. Preparation of mouse monoclonal antibodies.

only the gene of choice, but also the associated regulatory and expression sequences. Finding this segment requires mass cloning and a selection method to facilitate growth of the desired clone or a method to locate the proper clone among a large number of inappropriate ones. Examples of methods for clonal selection include gene-product-specific staining techniques or antimetabolite resistance.

## Monoclonal Antibody Production

The traditional methods for raising antibodies against a purified antigen involve injection of the antibody into a responsive animal (e.g., rabbit, goat, horse, etc.) and after a period of time collecting sera from the exposed animal. Antibodies raised against the antigen will be present in the sera and will react with the antigen, as evidenced by precipitation, neutralization, or other biochemical reaction. Sera containing high titers of antibody could then be used to detect the presence of the "stimulating" antigen in various biological samples. Generally, the antigen used to raise the antibodies would induce the production of a wide range of antibodies to itself (polyclonal). This range of antibodies often made specific antigen detection ambiguous due to cross-reaction of one or more of the antibodies in the polyclonal array to antigens other than the stimulating antigen. Figure 9.7 illustrates the important components of an antibody molecule.

Highly specific monoclonal antibodies (MoAb) are derived from hybrids of spleen B lymphocytes (antibody-producing) cells derived from a challenged animal and myeloma cells. The B lymphocytes respond individually to antigens by producing antibodies conforming (variable end of the molecule) to the antigen (Figure 9.7). Each B cell, once challenged by the antigen, is committed to the production of a single monospecific antibody against one aspect of the antigen for the duration of its existence. Myeloma cells are tumor cells that are immortal and capable of continuous production of antibodies. Several hybridomas to the antigen may be produced with various affinities and specificity for recognition. Selecting the most appropriate clone is often a laborious process. The resultant hybrid clone of the fusion of a specific committed B lymphocyte and myeloma cell is capable of manufacturing quantities of a single antibody uniquely keyed to the initiating antigen. This process is shown in Figure 9.8. Monoclonals are useful in that (1) they are monospecific and (2) they can be produced in bulk quantities as reagents. As will be described later, MoAbs have tremendous potential application to genetic toxicology.

## APPLICATION OF TECHNIQUES TO GENETIC TOXICOLOGY

The three techniques or products described previously are becoming increasingly important in developing better assays to detect and quantify genetic toxicity. The following summary gives only a brief indication of their potential applications.

### Assay Development

Figure 9.4 shows the application of recombinant DNA techniques to the development of new, more sensitive mutagen tester strains. Using similar methods, one can construct hybrid cell assys (e.g., a mammalian cell target gene in a prokaryotic cell host) to place relevant genes in host

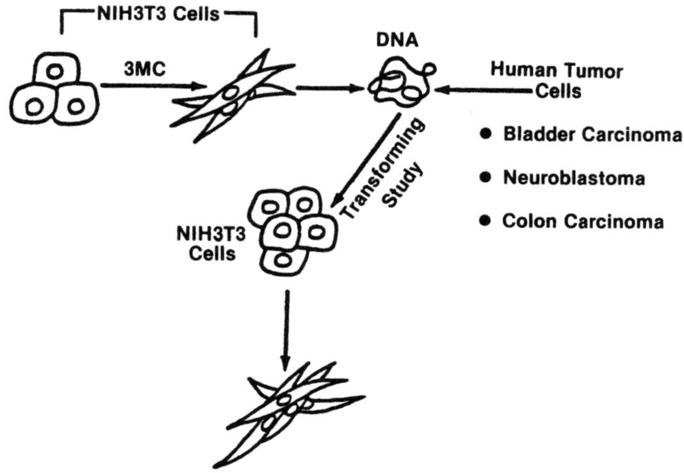


FIGURE 9.9. Indentification of transforming genes by transfection of isolated DNA from chemically transformed cells. These studies demonstrated the presence of discrete genetic sequences (oncogenes) responsible for cell transformation.

cell types that can be conveniently handled. For example, human proto-oncogenes might be cloned in a mammalian cell line more easily adapted to a routine tester organism than human cells.

MoAbs have been produced that recognize BrdU/DNA. Such monoclonals have been used as specific staining reagents for sister chromatid exchange (the BrdU-containing chromatid is recognized by the MoAb). Another application of MoAbs specific for BrdU/DNA might be to locate DNA repair patches formed during posttreatment incubation in BrdU-containing medium. Using either a radiolabel or fluorescent tag on the MoAb, short and/or long repair patches should be visible in cells exposed to genotoxic agents. Theoretically, this approach would be more specific and sensitive than current autoradiographic UDS methods.

A concern has been raised that protein pharmaceutical products produced from engineered cultured mammalian cells may carry contaminant DNA containing activated oncogenes. Chronic therapy with the drug might result in the uptake and expression of one of these genes by the user. Activation of cellular proto-oncogenes has been investigated using a technique in which purified DNA isolated from transformed cells is cocultivated with nontransformed target cells under conditions permitting direct DNA uptake and expression of activated oncogenes (see Figure 9.9). This process is called transfection and is described in detail in Chapter 8. DNA transfection is now conducted on a somewhat routine basis to screen tumor cell DNA for the presence of activated *ras*-group transforming oncogenes. The transfection assay coupled with a method to amplify the containing DNA might be a useful screen to assure the absence of biologically active contaminant DNA fragments, at least for the *ras* type of oncogene, and might well be employed as a quality control step for all forms of bioengineered products derived from cloned mammalian genes or made in mammalian host cells.

## Human Monitoring

Progress in the application of new molecular methods to human monitoring has already begun. One of the most rapidly developing and potentially most profitable applications of biotechnology involves the preparation of MoAbs to specific chemical–DNA adducts. The resulting MoAbs are then applied to human cells to quantify the level of DNA exposure to the chemical.

Adduct formation is a measure of chemical damage of cellular macromolecules; however, it must be emphasized that not all adducts are associated with adverse health effects. Genotoxins form different classes of adducts with proteins, RNA, and DNA. Only a few of the possible DNA adducts produced by a chemical may be involved with mutation. Consequently, estimation of risk would necessitate (1) the identification of the relevant DNA adduct or adducts and (2) a method that monitors for that specific adduct. Measurements of total DNA binding may be useful in dosimetry analysis, but not in toxicity production.

The most promising approach to the use of DNA adducts for either dosimetry studies or risk assessment is based on the immunoassay detection of specific DNA adducts in human circulatory blood cells. Using antibodies directed at specific DNA adducts, one can detect a single adduct in $10^7$–$10^8$ unmodified bases.[6]

The production of useful antibodies requires knowledge of the interactions between DNA bases and the active form of the chemical. Precise bonding relationships are necessary in order to construct base–chemical adducts that can be used as antigens to raise the antibodies. As shown in Figure 9.10, cis-platinum forms at least four types of DNA adducts. Each would have to be isolated, chemically characterized, synthesized, and conjugated for use as an antigen to raise MoAbs. This requires substantial capability in synthetic chemistry.

A second factor affecting this type of analysis is the rate of repair for different adducts. Some adducts, such as monofunctional DNA alkylations, are quickly repaired (i.e., most are eliminated within 24 hr after treatment) whereas others, such as inter- and intrastrand cross-links, are much more persistent. Thus, the repair rate will influence greatly the

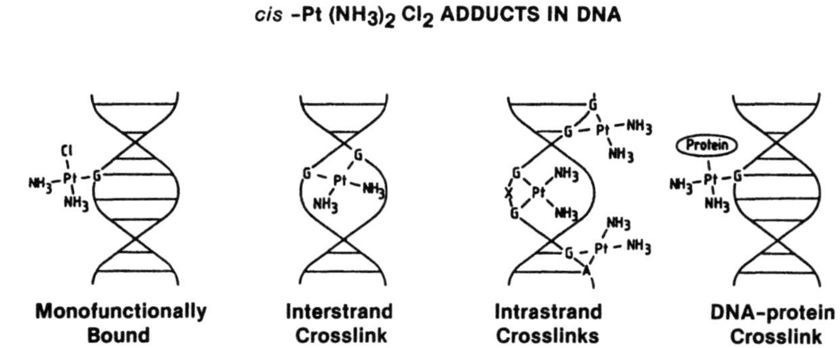


FIGURE 9.10. Cis-platinum, a cancer chemotherapeutic drug, interacts with DNA to form at least four specific types of DNA adduct. In order to produce MoAbs to each specific adduct, the cis-platinum–DNA complexes must be synthesized chemically and employed as antigens. (Adapted from Reedijk and Lohman.[4])

persistence and hence detection of adducts using immunochemical methods.

Use of adducts as a component of assessing genotoxic risk will ultimately necessitate a determination of which adducts lead to toxicity and mutagenicity and which are not toxicologically relevant.

In addition, the kinetics of various adducts may vary considerably; thus, measuring the efficiency of total DNA binding may lead to erroneous interpretations. For example, a hypothetical mutagenic carcinogen yields four different aklylation products upon administration to rodents. Among these four, those that alkylate at the 0–6 position of guanine and 0–4 position of thymine appear to be particularly important in producing point mutations. When analyzed *in vivo*, the total DNA binding of aklylation products of this hypothetical mutagen may increase with exposure level, leading to an assumption that genotoxic risk increases proportionally with dose. However, when the binding efficiencies of individual adducts are followed, the biologically relevant adducts could be produced at efficiency kinetics inversely related to dose, but masked by the formation of a large proportion of nonrelevant N-7 adducts of guanine (Figure 9.11). In this situation low dose rates may give proportionally greater increments of risk than high doses.

It is evident from this discussion that adduct analysis is potentially a powerful technique with a high degree of sensitivity and specificity. The limitations are that a great deal of information is needed about the array of adducts produced and their biological fate. For most chemicals, the effort to acquire such data is not justified and most of the work involving immunochemical staining of adducts in human cells to date has been with

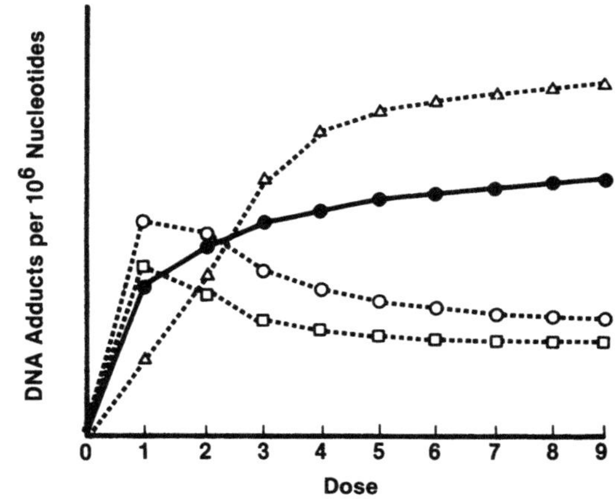


FIGURE 9.11. Evaluation of binding efficiencies may lead to erroneous conclusions about adduct–dose relationships unless analysis of each adduct is assessed independently. This figure illustrates how the induction kinetics of two biologically relevant adducts (○) $O^6$-ethylguanine and (□) $O^4$-ethylthymine can be missed by measuring (●) total DNA binding; (△) $N^7$-ethylguanine.

either extremely well-characterized and ubiquitous mutagens[10] or cancer chemotherapeutic drugs.[4] Use of monoclonal antibodies in combination with electronic amplification of the fluorescence signals and a computer-base image analysis has been successful in quantifying the adducts in single human lymphocytes, making the technique readily applicable to large-scale monitoring (see Chapter 6).

Nucleic acid hybridization with specific DNA probes also offers an opportunity for monitoring human cells. Proposals have been put forth to use DNA probes as diagnostic tools for cancer by detecting activated/mutant oncogenes and as diagnostic tools to identify genetic disorders via location of DNA lesions in specific genes. This latter application might work well in prenatal diagnosis. Amniocytes obtained from amniotic fluid can be cultured and fixed on slides. Chorionic cells are also obtainable from new fetal monitoring techniques. DNA probes for a wide range of mutations detectable in infants can be applied to individual cell preparation. Detection of gene sequences associated with increased susceptibility to specific pathologies may also be feasible as a consequence of the rapid rate of human gene mapping.

## Site-Specific Mutagenesis

Situations arise in which inactivation of a selected gene or portion of a gene would be desirable. Examples might be:

1. The need to specifically inactivate the gene controlling replication of a virus intended for use as a vector for rDNA. If viral replication would result in death of the infected host cell, use of the virus as a vector would be self-defeating. A mutation placed in the gene

that controls a required component of replication would prevent host cell death, yet leave other, desired genes intact.

2. Site-specific mutagenesis in genes producing a commercially valuable protein might improve the yield or biochemical or physical properties of the product (e.g., live vaccines).
3. Site-specific mutagenesis will enable scientists to determine the functions of each nucleotide segment of genes involved in biochemical pathways. Alteration of specific sites permits questions to be asked about the role of that gene.

The methods used to generate site-specific mutation involve restriction endonucleases of the type that make staggered cuts in defined sequences. The *Eco*RI restriction enzyme, for example, will cut between adjacent the G/A base sequence in one strand of DNA when ethidium bromide is present.

Once the cut is made, a gap can be generated by the use of an endonuclease followed by replacement of the excised bases with a mutagenic base analog. The thymidine analog *N*-hydroxycytosine is often used because it will hydrogen-bond with adenine or guanine, thus bringing about A–T to G–C transition. As more and more genes become available for cloning, site-specific mutagenesis will undoubtedly become a very valuable tool for assessment of gene function and gene manipulation.

### Restriction Fragment DNA Analysis

The many roles of restriction enzymes have been illustrated previously; however, polymorphisms found in restricting fragments from DNA of different sources can be used to provide general information regarding mutational changes.

Because restriction enzymes reduce large DNA polymers to smaller fragments by "cutting" the DNA at specific nucleotide sequences, the pattern of restriction fragments produced by a single enzyme should appear identical when separated on gel electrophoresis. If, however, alternatives in nucleotide sequences of the DNA have been introduced by genotoxic agents, the restriction fragments from a clone of treated cells will show a different pattern (polymorphism) in the stained gel. Polymorphisms indicate changes in the number or location of nucleotide sequences associated with restriction endonuclease excision points (i.e., mutations).

## ASSOCIATED IMPLICATIONS OF BIOTECHNOLOGY

### Gene Therapy

The impact of biotechnology on medical and toxicologic sciences has been limited to monitoring and diagnostic methods and to product de-

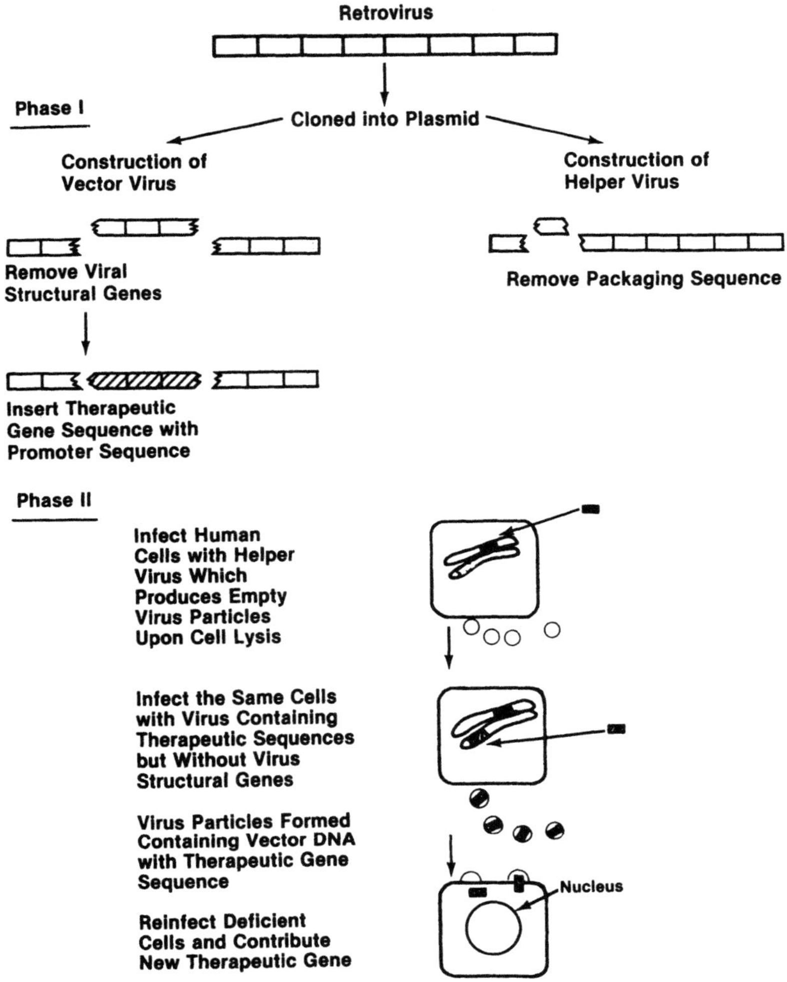


FIGURE 9.12. Retrovirus method of gene therapy.

velopment; however, as the abiliity to sequence and synthesize genes improves, the prospects for gene replacement grow as well.[11] In fact, introduction of new genes in laboratory animals has already occurred. A strain of mice with genetically produced reduction of growth hormone has been altered by the introduction of a rat growth hormone gene into the fertilized egg of an affected mouse. The rat DNA became incorporated into the mouse chromosome of some of the fertilized eggs and was expressed.[12] The new animal is designated as a transgenic strain. The de-

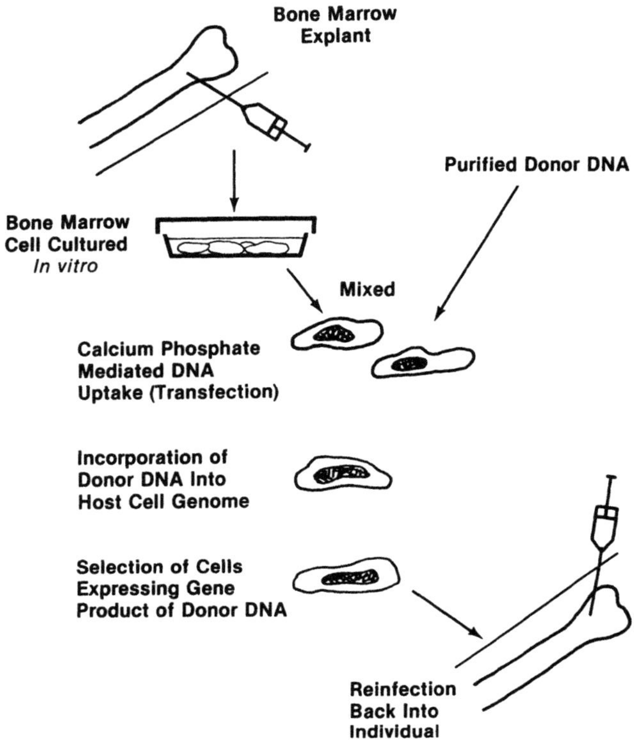


FIGURE 9.13. Bone marrow method of gene therapy.

velopment of transgenic animals containing defective genes might be a useful method of constructing better animal disease models.

Zygotic microinjection is one example of how new expressible genes can be introduced early into the developing mammalian embryo. It may be feasible to accomplish this type of gene therapy in human *in vitro* fertilization procedures, but the current success rate on a per-zygote-basis is low; therefore, other methods of introducing genes into adult organisms might be of greater near-term application. Two methods of choice are the use of retroviral vectors and bone marrow cell transplantation.

Methods proposed for retroviral gene therapy consist in the elimination of oncogenic properties associated with retrovirus followed by insertion of the therapeutic gene within the viral gene sequence in such a way that it is expressed once integrated into the host cell genome. The engineered vector contains enough viral genes to ensure that the therapeutic gene is carried into a few host cells placed in culture and then integrated and expressed. Clones of the cultured cells showing evidence of gene expression may then be isolated and reintroduced back into the host, where virus capsids containing the therapeutic gene will be formed and infect other cells, with the concomitant introduction of the functional gene (Figure 9.12).

Because the retrovirus insertion sites into the host cell genome cannot be predicted, it is conceivable that insertion might occur in a promoter sequence and activate that gene (e.g., oncogenes). Problems of this nature must be resolved prior to the use of this method.

Bone marrow replacement methods utilize the fact that marrow cells can be removed from the patient and grown temporarily in tissue culture. During tissue culture growth, new genes are introduced into the cells by one of several techniques (e.g., transfection), and those cells that have successfully acquired the gene are selected for reintroduction back into the host (Figure 9.13). The reason for selecting bone marrow cells, in addition to their potential for *in vitro* growth, is that bone marrow is a source of many enzymes and proteins found in blood. Consequently, blood proteins produced by the marrow cells will be delivered systemically by normal blood flow.

Selection of genetic diseases most likely to be high on a list of candidates for gene therapy are those that result from serum enzyme deficiences. A disease such as Lesch–Nyhan (HGPRT deficiency) is a good example. This deficiency could be repaired in adults via the bone marrow transplant approach. Table 9.1 summarizes methods applicable to gene introduction.

Somatic cell gene therapy will not affect the mutant load in the human gene pool. An individual treated for Lesch–Nyhan disease using the bone marrow implant technique will still produce gametes carrying HGPRT-deficient alleles. Gene replacement at the gamete or zygote level is the only method likely to result in reduced genetic load, since all somatic and germ cells will carry the new gene. The fact that some of the genetically repaired mice described earlier in this section proceeded to produce offspring with normal growth hormone levels shows that transplanted genes will progress through gametogenesis and fertilization processes.

The implications of gene therapy and gene improvement are staggering. While the concepts have been proven in principle, careful planning and decision-making are required to put them into practice.

## Oncogenes

### Discovery

The somatic mutation theory for cancer initiation suggested the presence of genes that, when genetically altered, would trigger cell transfor-

TABLE 9.1
Some Gene Therapy Techniques Being Considered for Human Applications[a]

| | Advantages | Disadvantages | How far advanced | How likely a candidate for use in humans |
|---|---|---|---|---|
| Chemical techniques (calcium phosphate-mediated DNA uptake and related techniques) | Easy and inexpensive to perform, requiring no special equipment<br>Do not require an infectious agent, so are very safe | Uptake and expression of exogenous DNA is low (for some cell types as low as one cell in one million; for the best, about one cell in 100)<br>Inserts multiple copies of the tranferred gene linked head to tail | Functioning genes that convey drug resistence (a stong selective advantage) have been transferred to mouse bone marrow cells; the cells replicate in the host mouse | Second most likely technique; low efficiency makes use in humans unlikely unless coupled with strong selective advantage, but apparent safety could make important if retroviruses turn out to be hazardous |
| Fusion techniques | Somewhat more efficient than chemical techniques<br>Do not require infectious agent, so are very safe | More complicated than chemical techniques | Have been used extensively in conventional tissue culture experiments, but not with cells in suspension, such as bone marrow cells | Similar to chemical techniques; however, unlikely to be used in preference to chemical techniques unless further work establishes them as in some way superior |
| DNA viruses | Effective and well-studied vectors for gene transfer in plants, animals and bacteria | Many of the most widely studied can be replicate within cells and some can kill the host cell<br>Less efficient as transformation than retroviruses | Relatively safe and efficient system using adenovirus recently has been reported on cells in culture | Hazard and low efficiency make human use unlikely at present |

| | | | | |
|---|---|---|---|---|
| Microinjection | Up to one in five injected cells carries the gene permanently | Must inject cells one at a time<br>Inserts are multiple copies of the transferred gene linked head to tail<br>Many cells are damaged during injection process<br>Probably feasible only for germ cell treatment | Genes transferred to fertilized mouse egg have been expressed in the mature mouse and transmitted to offspring in normal Mendelian manner; control of gene expression is not yet perfected | Inefficiency of technique and ethical questions about altering germ cells make near-term application unlikely |
| Retroviruses | Up to 100% of cells can be infected and express integrated genes<br>Can infect millions of cells at once<br>DNA integrates as a single copy at a single site<br>Retroviral vector does not normally replicate in host cell and does not appear to harm it | Can spontaneously rearrange and exchange sequences with other retroviruses, many of which are infectious, and some of which cause serious diseases (cancer and perhaps AIDS)<br>Retroviruses sometimes splice out inserted genes | Have been used to transfer functioning genes to bone marrow cells in mice; genes are expressed in the host for at least 6 months; control of expression is not yet perfected | Most likely technique so far; first human experiments probably will occur within the next 2 years |

[a] Adapted from Rawls.[13]

mation and tumor growth. The number, location, and function of such genes could only be speculated, upon, but studies on the kinetics of cell transformation by mutagenic carcinogens attempted to address this issue by developing mathematic models to calculate the number of genes involved in the transformation process. Most attempts were unsuccessful.

In the early and mid 1970s, the "carcinogens are mutagens" philosophy was extensively pursued through correlation studies involving short-term *in vitro* assays and rodent cancer bioassays.[14,15]

In the late 1970s, experiments were conducted that provided more direct support for the somatic mutation mechanism of cancer initiation. Shih *et al.* reported that DNA from carcinogen-transformed mouse fibroblasts was capable of transforming normal fibroblasts following direct DNA transfer (transfection) conditions (Figure 9.9).[16,17] They concluded that these results supported the existence of a transmissible transforming gene. Additional studies with restriction endonucleases suggested that the transforming gene is carried in a DNA fragment of approximately 20 kilobases. The responsible genes were identified as oncogenes.[18] The investigations assumed that 3-methylcholanthrene was directly responsible for the effects via mutagenesis. Studies with DNA obtained from tumor cell lines as well as tumor cells also showed positive transforming activity. The sources of these DNAs were mouse, rabbit, and human.

The fact that DNA codes for proteins that transform cells has been evident for many years from studies of oncogenic retroviruses in birds and nonhuman mammals. Retroviruses and RNA viruses produce effects by undergoing nonlytic infection of the host cell, producing reverse transcriptase, which allows the RNA to be copied into DNA and the new DNA sequences to become incorporated into the genome of the host cell. Proteins from these newly integrated genes stimulate the cells to go into a state of unregulated proliferation. The mechanisms by which retroviruses induce cellular proliferation have only recently been elucidated, and there appear to be two pathways.

The first mechanism involves the chance integration of the reverse-transcribed DNA sequences into a host cell oncogene. The insertion of the viral genome results in the transcription of the viral gene products as well as the oncogene products.

The second mechanism requires that the retrovirus acquire a cellular oncogene via a previous host cell lytic infection. The oncogene-containing retrovirus will carry the oncogene into a new host cell and, if it becomes integrated, the oncogene may be expressed, resulting in cell transformation.

Thus, it appears that even oncogenic viruses require an association with cellular oncogenes to function as cancer inducers.

Although it is widely presumed that human oncogenes are related or may have been acquired via retroviral mechanisms, cancer in humans appears to be minimally related to retrovirus injection (except for HTLV-III) and more so to chemical exposure. In fact, virtually all of the chemicals identified by the IARC as human carcinogens (Table 9.2) are active in one or more short-term tests for genotoxicity. Thus, one can hypothesize that direct genetic alteration of an oncogene by a chemical carcinogen may be the common mechanism for human cell transformation. Studies of alterations in the DNA sequences of some oncogenes (*ras*) obtained from human tumor cells show that point mutations are present in the DNA (Table 9.3).[22,23] Oncogene DNA from nontumor cells is free of these mutations. Chromosome breakage and gene translocations at the sites of some oncogenes (*myc*) are also associated with human neoplasms.[24] Figure 9.14 illustrates the locations of many of the known oncogenes on the chromosomes of the human genome.

Unfortunately, the problem of cancer in human populations appears to be more complex than exposure to genotoxic agents. For example, one group of chemicals, synthetic progesterones, is not genotoxic. Steroidal and hormonal compounds are included with other chemicals of diverse structures into a group of nongenotoxic carcinogens. This group has also been referred to as epigenetic carcinogens.[25] There appears to be no direct interaction such as covalent binding between these chemicals and DNA that might associate them with oncogene activation. A more complex mechanism may be required to explain the tumorigenicity of these chemicals. One mechanism that might be responsible for the majority of materials involves the interaction between the nongenotoxic carcinogen (e.g., progesterone) and a protein receptor required for expression of the biological activity of the hormone. Together, the hormone–receptor complex interacts with sites on DNA to initiate transcription of specific genes. Chronic exposure to unusually high levels of the nongenotoxic carcinogen might result in transcription hyperactivity. In the process, genes not intended for transcription might be inadvertantly activated. Oncogenes or genes that regulate oncogenes could be triggered by this mechanism, resulting in cell transformation.

Through a broad interpretation of oncogene activation, one can theoretically unify what appears to be unrelated mechanisms resulting from mutagenic, viral, and nongenotoxic carcinogens.

## Types of Oncogenes

Early transfection studies suggested the existence of more than one type of oncogene. It appeared that all human and subhuman mammalian cells contained a complement of oncogenes that in their nonactivated state produced substances necessary for normal cell proliferation and cell sur-

TABLE 9.2
The Genotoxic Profiles of 23 Human Carcinogens[a]

| Group 1 chemicals | *In vivo* tumor assays | | | | | Mammalian cell mutation assays | Cell transformation | Cytogenetics | | | Human cytogenetics |
|---|---|---|---|---|---|---|---|---|---|---|---|
| | Human | Mouse | Rat | Other | Ames | | | ABS | SCE | UDS | |
| 1. 4-Aminobiphenyl | + | + | + | + | + | ND | + | ND | ND | + | ? |
| 2. Analgesic mixtures containing phenacetin | + | – | + | | + | ND | ND | + | + | ND | ? |
| 3. Arsenic and arsenic compounds | + | – | – | | – | + | + | + | + | ND | + |
| 4. Asbestos | + | + | + | + | – | (+) | + | + | ND | ND | ? |
| 5. Azathioprine | + | (+) | (+) | | + | ND | ND | + | ND | ND | – |
| 6. Benzene | + | – | – | | – | – | – | + | ND | ND | + |
| 7. Benzidine | + | ND | + | + | + | + | + | + | + | + | + |
| 8. *N*,*N*-Bis(2-chloroethyl-2) naphthylamine (chlornaphazine) | + | + | + | | + | ND | + | ND | ND | ND | ? |
| 9. Bis(chloromethyl)ether and chloromethyl methyl ether | + | + | + | | + | ND | + | + | + | ND | + |

| | | | | | | | | | | | |
|---|---|---|---|---|---|---|---|---|---|---|---|
| 10. Myleran | + | (+) | – | | + | ND | + | + | + | ND | + |
| 11. Certain combined lymphoma chemotherapy (MOPP) | + | ND | ND | | ND | ND | ND | ND | ND | ND | + |
| 12. Chlorambucil | + | + | + | | + | ND | ND | + | ND | ND | + |
| 13. Chromium and chromium compounds | + | – | + | | + | + | + | + | + | + | + |
| 14. Conjugated estrogens | + | ND | – | | – | ND | ND | – | ND | ND | ? |
| 15. Cyclophosphamide | + | + | + | | + | + | + | + | + | ND | + |
| 16. Diethylstilbestrol | + | + | + | + | – | (+) | + | + | + | – | ? |
| 17. Malphalan | + | + | + | | + | + | + | + | + | + | + |
| 18. Methoxysalen plus UV light (PUVA) | + | + | ND | | + | + | + | + | + | ND | + |
| 19. Mustard gas | + | + | ND | | + | + | ND | + | ND | ND | ? |
| 20. 2-Naphthylamine | + | + | – | + | + | + | + | + | + | ND | ? |
| 21. Soots, tars, and oils | + | + | – | | + | + | + | + | + | + | ? |
| 22. Treosulphan | + | ND | ND | ND | ND | ND | ND | ND | ND | ND | ? |
| 23. Vinyl chloride | + | + | + | + | + | + | ND | + | + | ND | + |

[a] Data from Refs. 19 and 20. ND, No data available; parentheses indicate limited data.

TABLE 9.3
Point Mutations That Activate *ras* Oncogenes in Human Tumors[a]

| Oncogene | Tumor source | Point mutation | Amino acid change | Codon |
|---|---|---|---|---|
| H-*ras* | | | | |
| T24 | Bladder carcinoma | G → T | Gly → Val | 12 |
| Hs0578 | Mammary carcinoma | G → A | Gly → Asp | 12 |
| Hs242 | Lung carcinoma | A → T | Gly → Leu | 61 |
| N-*ras* | | | | |
| SK-N-SH | Neuroblastoma | C → A | Gln → Lys | 61 |
| SW-1271 | Lung carcinoma | A → G | Gln → Arg | 61 |
| K-*ras* | | | | |
| Calu-1 | Lung carcinoma | G → T | Gly → Cys | 12 |
| SW 480 | Colon carcinoma | G ← T | Gly → Val | 12 |

[a] Adapted from Aaronson and Tronick.[21]

face properties. With increased research interest in oncogene number, structure, and function during the past several years, considerable information has become available.

Oncogenes appear to fall operationally into several classes (Table 9.4). The classification scheme is based in part on amino acid homology of the oncogene products and similarity of cellular functions.

Class I oncogenes appear to be related to the synthesis of cell surface receptor proteins. The cell surface receptors are activated by growth factors produced by other cell types. An activated cell will proliferate as long as the second cell type produces its growth factor in response to some external stimulus. Abnormal cells lose responsiveness to the presence or absence of the regulatory growth factor and continue to proliferate.

Class II oncogenes are proteins that have a common characteristic of regulating cellular metabolism.

Class III are oncogenes regulating nuclear activities, possibly cell cycling.

Class IV is a collection of relatively unrelated oncogenes, some of which produce cell growth factor proteins.

Analysis of oncogene mechanisms that either produce growth factors (*sis* oncogene and platelet-derived growth factor, PDGF) or receptors for growth factors (*erb*-B oncogene and epidermal growth factor, EGF) indicate that alterations exist in cells such that the growth factor is altered to stimulate cells without the necessary receptor, or the receptors are altered to an extent that they send out signals to stimulate cell growth in the absence of a growth factor. Detailed understanding of just how these

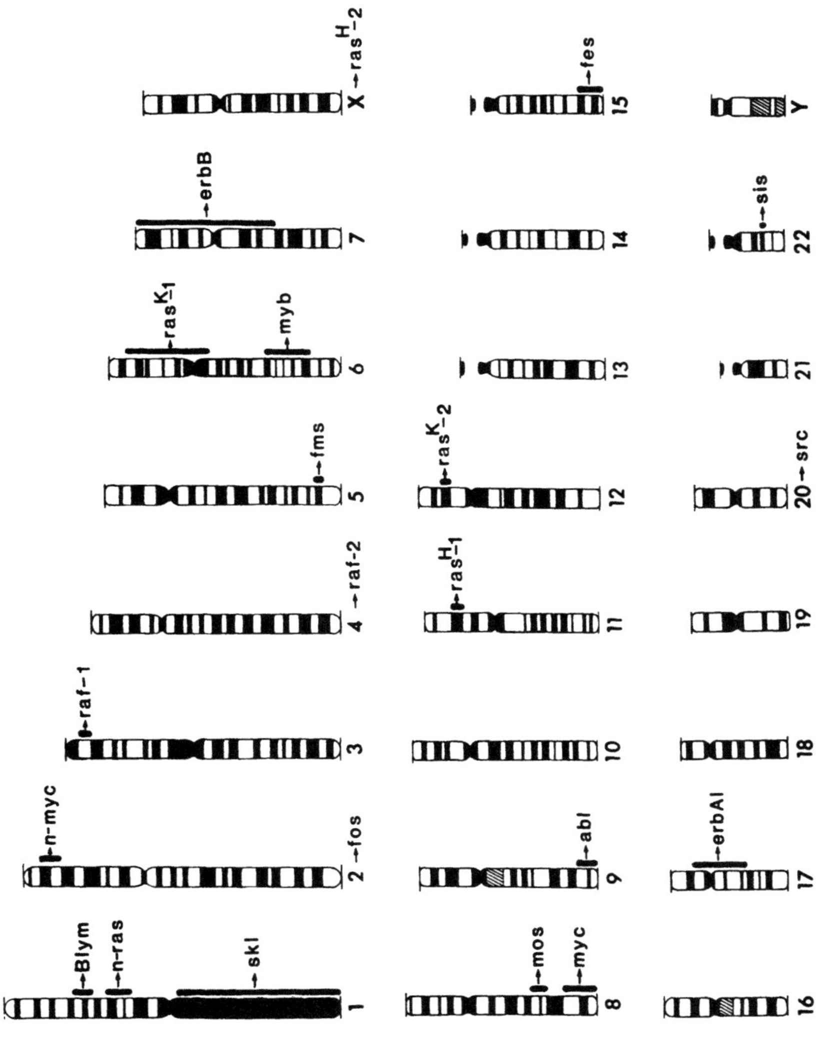


FIGURE 9.14. Schematic of human karyotype showing approximate location of known proto-oncogenes.

TABLE 9.4
Categories of Human Oncogenes

| Oncogene | Gene product characteristics |
|---|---|
| Class I (*src*, *yes*, *neu*, *abl*, *fps*, *fms*, *erb*-E, *ros*, *mos*, *fgr*) | Have extensive homology<br>Found in cytoplasm<br>Loosely bound to cell membrane<br>Have kinase (often tyrosine kinase) activity |
| Class II (HA-*ras*, $k_1$-*ras*, N-*ras*) | Have some homology<br>Loosely bound to cell membrane<br>Many bind GTP or GDP<br>Most commonly found oncogenes in human tumors |
| Class III (*myc*, *myb*, *fos*, N-*myc*, *ski*, p53) | Found in cell nucleus<br>Little protein homology |
| Class IV (*sis*, *rel*, B-*lym*, *erb*-A, *ets*, *met*) | No clear pattern |

relationships initiate transformation and tumor development is still several years away, but with the ability to sequence DNA, detect gene products, and clone oncogenes, the technology is available to provide eventual understanding of the intricacies of oncogene activity.

What has been clearly demonstrated, however, is that genetic damage directly to the oncogene structural DNA sequence (*ras*) or to its regulatory control (e.g., *myc*) will set in motion processes leading to cell transformation, cell proliferation, and ultimately metastasis.

The relationship of oncogenes to spontaneous tumorigenesis is also the subject of intensive investigation. Recent reports have indicated that resistance to chemotherapy might depend on the specific activity and level of oncogene product produced by an individual's cells. Thus, individual susceptibility to cancer and responsiveness to treatment may depend upon normal variability in gene expression.

## REFERENCES


1. Olson, S.: *Biotechnology, An Industry Comes of Age,* National Academy Press, Washington, D. C., 1986.
2. Office of Science and Technology Policy. Proposal for a coordinated framework for regulation of biotechnology. *Fed. Reg.* **49**:50856–50907, 1984.
3. Waife, S. O., and Lasagna, L.: From DNA to NDA—The impact of recombinant DNA technology on new drug development. *Regul. Toxicol. Pharmacol.* **5**:212–224, 1985.

4. Reedijk, J., and Lohman, P. H. M.: Cis-platinum; synthesis, antitumor activity and mechanisms of action. *Pharm. Weekbl. (Sci.)* 7(5):173–180, 1985.
5. Strudenski, A. B., Conner, B. J., Impraim, C. C., Teplitz, R. L., and Wallace, R. B.: Discrimination among the human $\beta^A$, $\beta^S$, $\beta^C$-globin genes using allele-specific oligonucleotide hybridization probes. *Am. J. Hum. Genet.* **37**:42–51, 1985.
6. Mekler, P. H., Delehanty, J. T., Lohman, P. H. M., Brouwer, J., Putte, P. v. D., Pearson, P., Pomwels, P. H., and Ramel, C.: The use of DNA technology to study gene alterations and gene evolution. *Mutat. Res.* **153**:13–55, 1985.
7. Maron, D. M., and Ames, B. N.: Revised methods for the *Salmonella* mutagenicity test. *Mutat. Res.* **113**:73–215, 1983.
8. Tindall, K. R., Stankowski, Jr., L. F., Machanoff, R., and Hsie, A. W.: Detection of deletion mutations in pSV2*gpt*-transformed cells. *Mol. Cell. Biol.* **4**(7):1411–1415, 1984.
9. Barton, K. A., and Brill, W. J.: Prospects in plant genetic engineering. In *Biotechnology and Biological Frontiers* (P. H. Abelson, ed.), American Association for the Advancement of Science, Washington, D. C., pp. 121–131, 1984.
10. Baan, R. A., Lansbergen, M. J., deBruin, P. A. F., Willems, M. I., and Lohman, P. H. M.: The organ-specific induction of DNA adducts in 2-acetylaminofluorene-treated rats, studied by means of a sensitive immunochemical method. *Mutat. Res.* **150**:23–32, 1985.
11. McCormick, D.: Human gene therapy: The first round. *Biotechnology* **3**:689–693, 1985.
12. Palmiter, R. D., Brinster, R. L., Hammer, R. E., Trumbauer, M. E., Rosenfeld, M. G., Birnberg, N. C., and Evans, R. M.: Dramatic growth of mice that develop from eggs microinjected with metallothionein-growth hormone fusion genes, *Nature* **300**:611–615, 1982.
13. Rawls, R.: Gene therapy. *Chemical and Engineering News* **62**:40, 1984.
14. Brusick, D. J.: Mutagenicity and carcinogenicity correlations between bacteria and rodents. In *Cellular Systems for Toxicity Testing 407* (G. M. Williams, V. C. Dunkel, and V. A. Ray, eds.), New York Academy of Sciences, New York, pp. 164–176, 1983.
15. Brusick, D. J.: The role of short-term testing in carcinogen detection. *Chemosphere* **5**:403–417, 1978.
16. Shih, C., Shilo, B., Goldfarb, M. P., Dannenberg, A., and Weinberg, R. A.: Passage of phenotypes of chemically transformed cells via transfection of DNA and chromatin. *Proc. Natl. Acad. Sci. USA* **76**:5714–5718, 1979.
17. Weinberg, R. A.: Use of transfection to analyze genetic information and malignant transformation. *Biochim. Biophys. Acta* **651**(1):25–35, 1981.
18. Bishop, J. M.: Cellular oncogenes and retroviruses. *Annu. Rev. Biochem.* **52**:301–354, 1983.
19. Evaluation of the Carcinogenic Risk of Chemicals to Humans. *IARC* October, 1982; suppl 4.
20. Garrett, N. E., Stock, H. F., Gross, M. R., and Waters, M. D.: An analysis of the spectra of genetic activity produced by known or suspected human carcinogens. *Mutat. Res.* **134**:89–111, 1984.
21. Aaronson, S. A., and Tronick, S. R.: Oncogenes. In *Medical Genetics: 1984*, Proceedings of symposium presented by the Foundation for Advanced Education in the Sciences, May 10–12, 1984, pp. 74–95.
22. Reddy, E. P., Reynolds, R. K., Santos, E., and Barbacid, M.: A point mutation is responsible for the acquisition of transforming properties by the T24 human bladder carcinoma oncogene. *Nature* **300**:145–152, 1982.
23. Varmus, H. E.: Mutations of cellular oncogenes as a basis for neoplastic change. In *Mutation, Cancer and Malformation* (E. H. Y. Chu and W. M. Generoso, eds.), Plenum Press, New York, 1984, pp. 61–77.

24. Rowley, J. D.: Human oncogene locations and chromosome aberrations. *Nature* **301**:290–291, 1983.
25. Weisburger, J. H., and Williams, G. M.: The distinct health risk analyses required for genotoxic carcinogens and promoting agents. *Environ. Health Perspect.* **50**:233–245, 1983.

CHAPTER 10

# Genetic Toxicology and Its Role in the Study of Congenital Malformations

## INTRODUCTION

Associated with genotoxic risks is a large body of toxic effects, which include congenital malformations, teratospermia and reduced fertility, spontaneous abortion, miscarriage, and stillbirth, and increased predisposition to various toxic and/or infectious agents. These effects are very difficult to incorporate into genetic hazard assessments because they include traits with irregular, and thus nonpredictable, inheritance patterns as well as self-limiting traits, such as spontaneous abortion, which cannot always be linked directly to genotoxicity.

These dysfunctions also are omitted as endpoints of genotoxic exposure since, while they may be the consequence of genetic damage to somatic and/or germ cells, it is also possible to provide a nongenotoxic origin for each (Table 10.1).

## CONGENITAL MALFORMATION RISK

### Origin of Congenital Malformations (Birth Defects)

The precise relationship between congenital malformations in mammals and mutagenic events remains a subject of investigation. As with the mutagenic load, an accurate knowledge of the malformation incidence

TABLE 10.1
Possible Origins for Reproductive Deficiencies and Malformations in Mammals

| Phenomenon | Possible nongenotoxic origin | Possible genotoxic origin |
|---|---|---|
| Congenital malformations | Teratogenic effect developed from maternal exposure during embryogenesis | Mutation induced in allele of a mutagenic trait<br>Somatic dominant mutation induced in the embryonic cell mass during middevelopment |
| Spontaneous abortion miscarriage, stillbirth | Detrimental nutritional, health, and emotional status of the mother during fetal development<br>Fetotoxic exposure of the mother during embryogenesis | Dominant mutation or chromosome aberration induced in egg or sperm prior to fertilization<br>Somatic dominant mutation or chromosome aberration induced during early development |
| Impaired fertility | Gametotoxic effects of agents or metabolites | Genotoxicity-induced functional sperm abnormalities |
| Enhanced predisposition to toxic or infectious agents | Tissue or organ deposition and storage of toxic or immunoregulator, which is activated at a later time due to weight loss, hormonal imbalance, or other homeostatic change | Mutation induced in an allele affecting chemical metabolism (e.g., AHH locus) or immune competence (e.g., macrophage activation), which would not be detected as a genetic disease or syndrome |

for a given population is essential in order to perform risk analyses. This problem appears to be more difficult in some way for congenital malformations, since incidences of malformation can vary considerably between populations. Estimates range from a frequency of 5.5 per 1000 births in Vietnam (1960–1970) to 17.3 per 1000 births in England (1960). These frequencies are derived from assessments at birth. Some malformations (e.g., deafness, urogenital disorders) are not diagnosed until later in life. Consequently, the congenital malformation incidences are usually revised upward in a followup assessment. In England, for example, the indicated incidence of 17.3 increased to 23.1 per 1000 births at 5 years of age.[1]

The source of human congenital malformations is complex and, as indicated in Table 10.1, includes genetic and nongenetic mechanisms. To put their origin into perspective with other types of genetically related phenomenon, three broad categories have been proposed for all human diseases/abnormalites; (1) diseases/abnormalities whose principal cause is an external factor, such as viral infection, chemical agent, or dietary deficiency; (2) chromosome aberrations and monogenic traits resulting from dominant, recessive, or X-linked mutant alleles; and (3) multifactorial traits resulting from adverse gene combinations and associated genetic predisposition to the diseases/abnormalities. Category 3 may be described differently by investigators (e.g., unknown causation, complex causation, etc.), but family studies support the assumption that multifactorial genetic causation or predisposition is acting in this category.

Attributing a congenital malformation in a specific individual to a specific cause is difficult, if not impossible, since many anomalies may be caused by mechanisms in all three categories. Deafness, for example, occurs as a result of dominant mutation, multifactorial recessive traits, and by external nongenotoxic exposures to the pregnant mother. Heart and eye defects are also examples of traits that have a common appearance but may have multiple causes.

Based on a number of epidemiologic studies, congenital malformations, as recognized at birth, occur with a frequency of approximately 15 per 1000 births.[1] An important question is the proportion of this recurrent population load that is due to each of the three categories described previously. A model illustrating the possible composition of recurrent congenital malformations in the human species is shown in Figure 10.1. The frequencies in this model are only rough estimates and have not been verified. The primary reason for constructing this model is to address issues specifically related to malformation risk estimates as part of mutagenic hazard assessments.

Several conclusions are suggested by this model:

1. Most of the congenital malformations in the human population (~90%) are the result of preexisting mutant germ cells or newly arising somatic mutations.
2. Increases in the incidence of congenital malformations may result from either maternal or paternal exposure.
3. Temporary clusters of congenital malformations resulting from maternal exposure to nongenetic teratogens will not alter the malformation frequencies derived on large population samples over time.
4. Maternal and paternal exposure to mutagenic agents capable of inducing germ cell effects may result in increased incidences of recurrent congenital malformations. Mutagens limited to somatic cell effects would act in an analogous manner to nongenetic teratogens with respect to long-term changes in malformation frequencies.

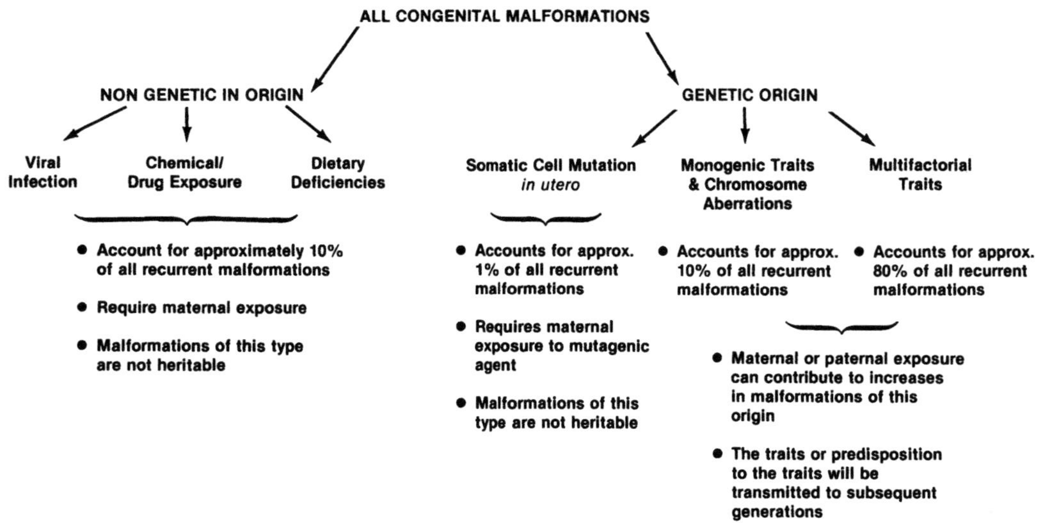


FIGURE 10.1. A model to define the composition and origin of congenital malformations in humans (estimated to be 15 per 1000 births based on assessment at time of birth or shortly thereafter; miscarriage, spontaneous abortion, and stillbirth are not included.

While clusters of malformation occur in humans as the result of maternal exposure to environmental agents (e.g., thalidomide, rubella, etc.), careful epidemiologic analysis often reveals the causative agent. Such nongenetic malformation clusters are easily identifiable due to the morphologic similarities in the malformation produced in affected individuals, such as phocomelia (thalidomide), and heart abnormalities and blindness in rubella exposure. Once a cause-and-effect relationship has been established, intervention can be established by prevention of exposure to pregnant women.

Genetically based (germ cell) malformations will occur randomly across all classes of anomalies and will not form a sentinel pattern or cluster. Consequently, causative information is almost impossible to obtain.

## Teratogenicity and the Male

An important concern has arisen during the past few years about male contribution to teratogenicity. Obviously, as indicated in Table 10.1, the male can theoretically contribute to genetically derived congenital malformations, reproductive deficiencies, spontaneous abortion, and other effects, but the issue of male contribution to nongenetic teratogenicity (such as raised by Agent Orange exposure to military personnel in Vietnam) requires a specific type of analysis.

Fetal malformations and morphologic aberrations can appear in many forms and exhibit a wide range of teratogenic significance. A report by Khera[2] proposed a classification scheme to organize different forms of anatomic changes and their significance (Figure 10.2). According to this scheme, malformations constitute the critical observations in teratogen-

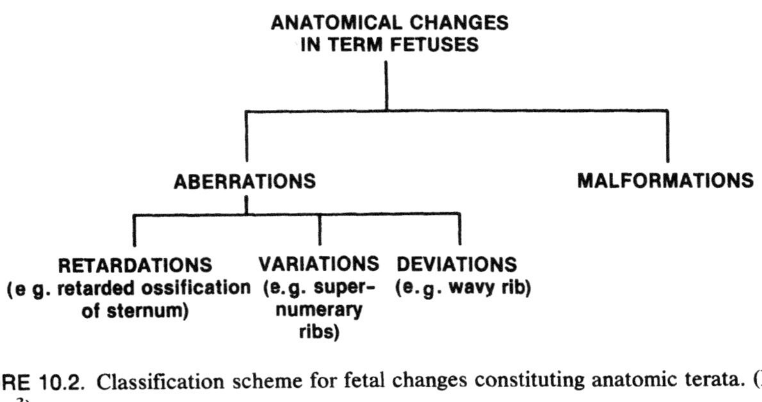


FIGURE 10.2. Classification scheme for fetal changes constituting anatomic terata. (From Khera.[2])

TABLE 10.2
The Absence of Male-Mediated Teratogenicity from Four Agents Known to Be Teratogenic in Females[a]

| Known female teratogen | Species | Genotoxicity status | Results of male-mediated evaluation |
|---|---|---|---|
| Thalidomide | Rabbits | Nongenotoxic | Reduced conception rate, but no significant increase in malformations in over 300 pups sired |
| Ethanol | Mice/rats | Very limited evidence for genotoxicity | Decreased litter size and reduced weight of offspring; no increase in malformations |
| TCDD (dioxin) | Rats | Nongenotoxic | Toxicity in males and some reduction in litter sizes and birth rates; no significant increases obtained in malformations |
| Lead | Rats | Uncertain, but generally considered nongenotoxic | Reduced litter sizes and pup weights; neonatal mortality increase, but no increase in malformations |

[a] Adapted from Pearn.[4]

icity assessment, with aberrations acting as contributing or confirmatory evidence. Aberrations alone would not be considered as more than a suspicion for teratogenicity.

Human studies focused on military personnel serving in Vietnam and presumably exposed to dioxins via Agent Orange emphasized the difficulties in using an epidemiologic approach to address this problem.[3] The conclusions reached following years of analysis suggest no anomalies in birth as a consequence of military service in Vietnam. However, for many these conclusions were insufficient to eliminate concern.

Experimental information that is available to answer the question of male-mediated teratogenicity is limited, but is consistent with the conclusions of the human studies. The approach involves performing teratogenic analyses of litters from untreated females mated to treated males. The males in this study would be exposed, ideally, with known teratogenic agents devoid of genotoxic effects in order to eliminate possible mutagenic contributions.

A series of studies generally meeting these criteria was reported by Pearn.[4] The results of Pearn's review are summarized in Table 10.2. While this list by no means resolves the issue, it supports the presumption that teratogenic syndromes cannot be induced in the offspring of exposed male mammals and that nonmutagenic teratogens are only effective when administered to the female at specific times during embryogenesis.

## SUMMARY

Congenital malformations make up a substantial portion of all diseases identified in human newborns. While factoring this information into genetic hazard assessments is complicated by the complex origin of such abnormalities, new mutations induced in recessive loci by genotoxins can contribute to a rise in this category of genetic disease.

## REFERENCES

1. Matsunaga, E.: Perspectives in mutation epidemiology. 1. Incidence and prevalence of genetic disease (excluding chromosomal aberrations) in human populations. *Mutat. Res.* **99**:95–128, 1982.
2. Khera, K. S.: Common fetal aberrations and their teratogenic significance: A review. *Fund. Appl. Toxicol.* **1**:13–18, 1981.
3. *Australian Royal Commission Report on the Use and Effects of Chemical Agents on Australian Personnel in Vietnam,* Volume 3: *Birth Anomalies*, Australian Government Publishing Service, Canberra, 1985.
4. Pearn, J. H.: Teratogens and the male: An analysis with special reference to herbicide exposure. *Med. J. Aust.* **2**:16–20, 1983.

# Appendixes

## APPENDIX A: S9 HOMOGENATE AND S9 MIX

### Background

*In vitro* metabolic activation of xenobiotics via hepatic S9 in microsomal supernatant was one of the early developments of genetic toxicology that stimulated the expansion of this field. Method development would be severely limited without some form of exogenous metabolism system that can be used to supplement the endogenous capacity of commonly employed target cells.

Prior to the application of a microsomal supernatant reaction mixture to bacterial mutagenicity screening,[1] more cumbersome methods, such as the host-mediated assay,[2] were used. The host-mediated assay utilized a host mammal such as the mouse to convert promutagens into their active metabolites. The bacteria were implanted in the mouse via intraperitoneal[2] or intravenous[3] injection. Application of the test material followed shortly after the bacteria, using a different route of administration. After an incubation period of 1 hr or so, the host animal was killed and as many bacteria as possible were recovered and plated onto selective media for mutant enumeration. The host-mediated assay had an appeal due to its relevance to hazard assessment, but unfortunately worked well only for a limited number of promutagens.[2,3,4] Consequently, S9 quickly became the basic activation method for routine testing.[5] In the mid-1970s, S9 activation technology was transferred to *in vitro* mammalian cell culture techniques.[6] Since 1975, with the publication by Ames and colleagues of the "standard" S9 mix, only minor changes have occurred in the production, quality control evaluation, and use of S9 mixes. A number of studies were conducted and published showing that useful S9 mix could be generated from organs other than the liver, from species other than the rat,[7,8] and induced with agents other than Arochlor 1254.[9,10] Still, hepatic S9 mix from the male rat is by far the most widely used S9 activating system.

How much do we know about this system? The S9 mix has been used virtually unchanged for more than 10 years, during which time data have been produced that affect the economic rate of industrial chemicals, pharmaceuticals, and other consumer products. The following are some questions commonly raised about S9 for which little information has been published.

1. How much variability can be expected from batch to batch of S9 homogenate?
2. Should S9 homogenate be standardized, and, if so, on what basis?
3. Can frozen S9 homogenate be thawed and refrozen without significant loss of activating potential?
4. What is the best method to store S9 homogenate?
5. How long will S9 mix (homogenate plus cofactors) remain active on ice during a work day?

These are important operational questions, which influence how laboratories operate. Other questions concerning the optimization of S9 mix levels for an assay, composition of cofactors, and timing of Arochlor 1254 induction have been studied more extensively and are generally provided in reports from the various EPA Gene-Tox working groups. Some activation modifications for specific types of chemicals, such as nitroamines and azo dyes, have become reasonably standardized.[11,12]

## S9 Batch Variability

Two biochemical parameters generally measured for S9 homogenate are mg/ml of protein and aryl hydrocarbon hydroxylase (AHH) activity. These two values plus a bioactivation titration of S9 with a *Salmonella typhimurium* revertant (e.g., TA98) and a fixed concentration of promutagen (e.g., benzo[*a*]pyrene) commonly constitute the quality control steps imposed on S9 production. Homogenate prepared from noninduced rat liver contains about 40 mg/ml of protein (Lowery method) and has AHH specific activity of 0.25 (nmole of substrate metabolized/mg protein/20 min reaction). Table A.1 gives the protein levels and AHH specific activity of 31 individual batches of S9 homogenate prepared over 24 months from Sprague-Dawley male rats.

TABLE A.1
Variability in Protein Content and AHH Activity in Different Lots of Arochlor-Treated Rat S9

| Lot no. | Date vialed | Protein (mg/ml) | AHH Activity (ng/20 min/mg protein) |
|---|---|---|---|
| CB024 | 02/07/79 | 39.5 | 3.77 |
| CB031 | 02/07/79 | 38.5 | 5.53 |
| CD083 | 04/19/79 | 24.8 | 0.48 |
| CE097 | 05/03/79 | 29.2 | 1.52 |
| CE098 | 05/31/79 | 27.9 | 1.80 |
| CE099 | 05/31/79 | 28.5 | 2.10 |
| CH147 | 09/07/79 | 37.4 | 1.20 |
| CI159 | 09/28/79 | 35.0 | 0.96 |
| CJ167 | 10/05/79 | 33.4 | 0.82 |
| CJ170 | 10/26/79 | 32.9 | 10.80 |
| CH172 | 11/09/79 | 27.0 | 9.97 |
| CH173 | 11/09/79 | 26.3 | 7.50 |
| CH177 | 11/27/79 | 40.0 | 13.93 |
| CH180[a] | 12/10/79 | 40.0 | 11.50 |
| CI182 | 12/10/79 | 40.0 | 15.00 |
| DA006 | 01/25/80 | 29.3 | 15.56 |
| DB009 | 02/22/80 | 33.0 | 21.93 |
| DE035 | 05/12/80 | 37.6 | 7.26 |
| DE036 | 05/22/80 | 35.1 | 8.95 |
| DF063 | 07/02/80 | 41.0 | 3.57 |
| DF064 | 07/02/80 | 36.0 | 1.58 |
| DF065[a] | 07/02/80 | 56.0 | 4.44 |
| DG074 | 07/21/80 | 26.5 | 6.40 |
| RDH088 | 08/11/80 | 29.6 | 6.37 |
| RDH092 | 08/29/80 | 31.8 | 5.05 |
| RDH093[a] | 08/29/80 | 25.1 | 6.03 |
| RDI095 | 09/26/80 | 34.0 | 1.19 |
| RDJ108 | 10/10/80 | 19.5 | 10.84 |
| RDJ109 | 10/13/80 | 30.0 | 1.26 |
| RDJ112 | 10/20/80 | 19.0 | 9.41 |
| RDL127 | 01/09/81 | 24.5 | 8.80 |

[a] Prepared in sucrose rather than KCl.

The induction procedures with Arochlor 1254 and all manufacturing steps were standardized as much as possible.

There does not appear to be any clear relationship between protein content and AHH activity (Figure A.1). Biological activity is also variable among batches, and in this case there does appear to be a slight positive relationship between AHH specific activity and activation of polyaromatic promutagens.

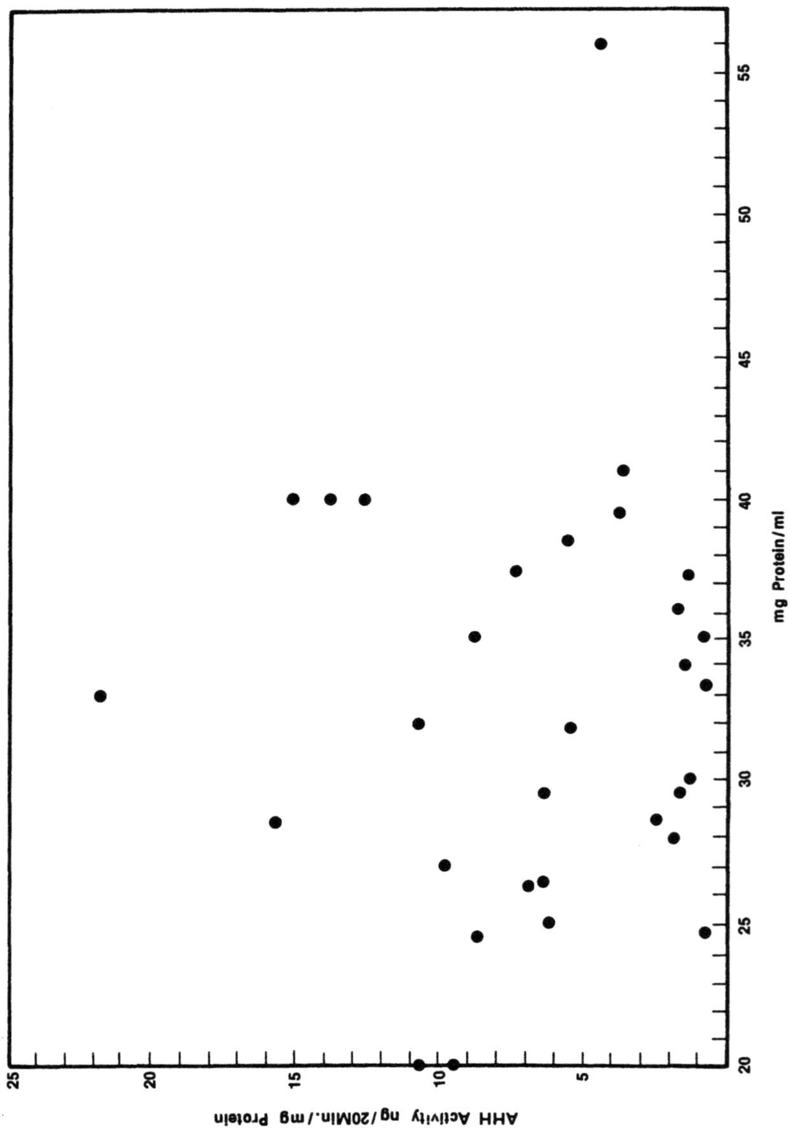


FIGURE A.1. The relationship between S9 homogenate protein and aryl hydrocarbon hydroxylase (AHH) activity.

## S9 Standardization

The extent of S9 variability described previously would argue for some sort of standardization. However, it is unclear as to what parameter might be the most useful for standardization. A biochemical measurement such as mg/ml of protein would be a simple and quantitative measure, but it appears that protein content is not predictive of biological activity. AHH activity might be useful, but this provides insight into only one (p448) of the many mixed-function oxidase enzyme complexes. Bioassay against one or two reference promutagens may be suitable, but such an approach might only optimize S9 for one or two chemical classes in a single bioassay. Reference standards for easily measured parameters do appear to be the appropriate approach. Acceptable contamination levels in S9 products should also be included in standardization criteria. One possible approach might be to establish a standard reference agent (or mixture of agents) with standard mutation induction curve developed and published for use in standardizing different lots of S9 with specific mutagens.

## S9 Stability

Stability of specific enzymes in frozen S9 has been analyzed and shows that some are very stable and others are not.[13] For routine testing of chemicals where the responsible enzymes for biotransformation are seldom known, this information may not be extremely helpful. However, some examples may be useful as surrogates for evaluating laboratory handling procedures. For example, Table A.2 shows the effects that thawing and refreezing of S9 have on mutagenic activity of two promutagens. The effects appear to be minimal. Figure A.2 shows the change in bioac-

TABLE A.2
Activity of Aroclor 1254-Induced Rat Liver S9 in Activating Mutagens Following Freeze–Thawing

| Number of times freeze–thawed | Mutagenic activity[a] | |
|---|---|---|
| | TA98 | TA100 |
| 1 | 554 (BaP) | 2385 (2-AA) |
| 2 | 533 (BaP) | 2250 (2-AA) |
| 3 | 506 (BaP) | 2053 (2-AA) |

[a] Activation potential at 100 μl S9. BaP, Benzo(*a*)pyrene; 10 μg/plate; TA98 spontaneous background was 38. 2-AA, 2-Aminoanthracene; 2.5 μg/plate; TA100 spontaneous background was 185.

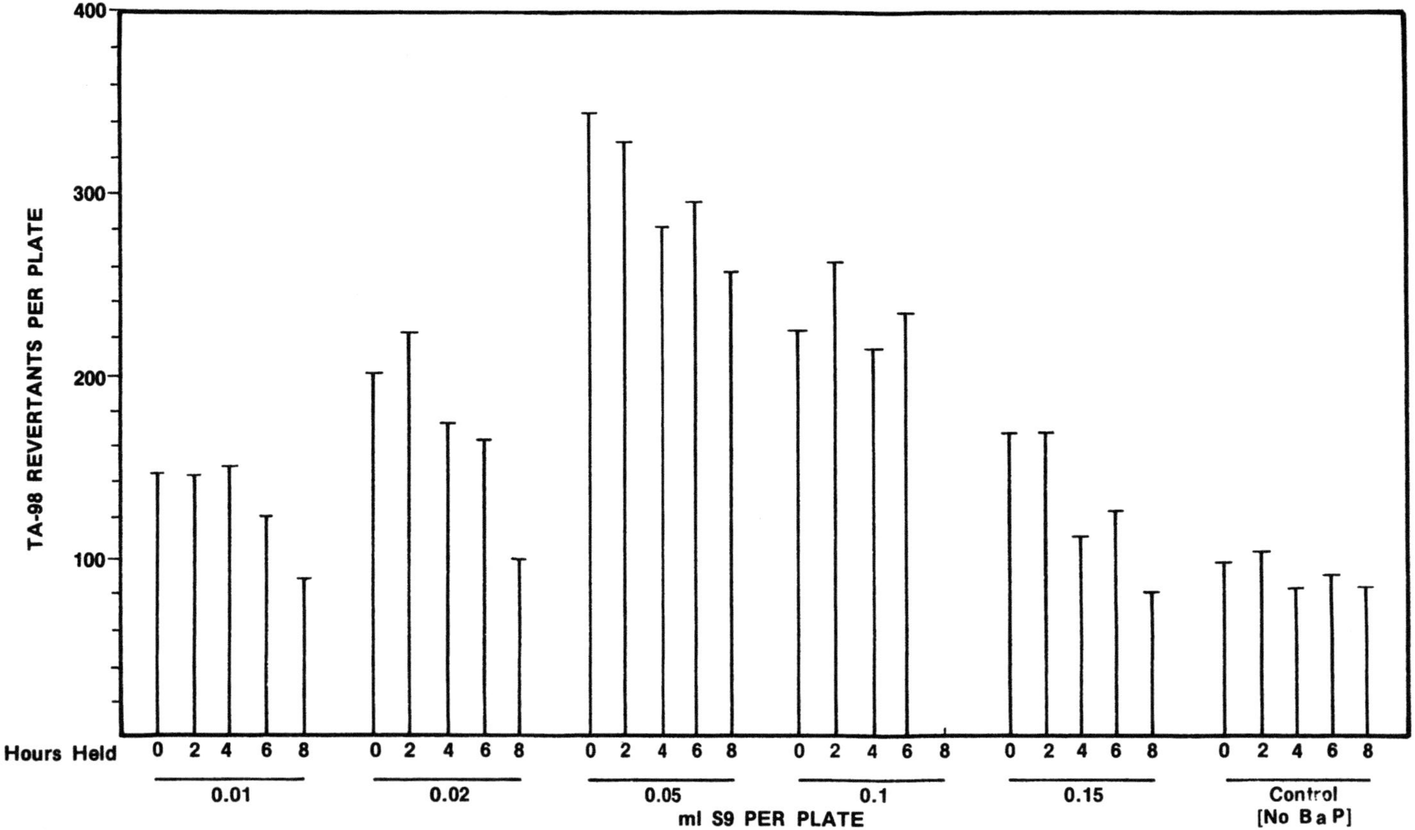


FIGURE A.2. Loss of S9 activity over 8 hr maintained at 0°C. Results are average of two plate counts for 5 μg benzo[*a*]pyrene (BaP) per plate using *Salmonella* strain TA-98 and S9 from Aroclor 1254-induced rats (IRL-78).

tivity of S9 mix held in an ice bath over an 8-hr period. These data are useful in knowing how long S9 mix might be held in the laboratory before significant deterioration occurs. Based on Figure A.2, one would limit laboratory holding of S9 mix to no longer than 4 hr in an ice bath. The stability of S9 mix in overlay agar has been reported by Nohmi *et al.*[14]

## S9 Storage

Most investigators use −80°C low-temperature freezers or liquid nitrogen to store S9 homogenate. Both methods appear to be adequate for up to 12 months. Beyond that period, we have little information. Undoubtedly, even in liquid nitrogen some breakdown occurs after extended storage. As long as stored S9 is used within 6–12 months, no serious problems appear to be present.

Longer storage may be possible using lyophilization in the production of S9 homogenate. Currently, two manufacturers of test kits for genotoxicity testing employ lyophilized S9. Tables A.3–A.5 show a comparison of frozen and lyophilized S9 using the standard Ames reverse mutation assay and a preincubation modification. The two S9 products appear to give equivalent activity. Obviously, lyophilized S9 has many potential advantages, particularly regarding shelf-life. Limitations appear to involve reconstitution of the lyophilized material. It tends to clump, rather than disperse evenly, in aqueous reconstitution solutions; however, this problem should be manageable through the addition of surfactants.

TABLE A.3
Titration of Lyophilized and Frozen Aroclor-Induced Rat Liver S9 against Benzo(*a*)Pyrene, 2-Acetylaminofluorene, and 2-Aminoanthracene Using TA100[a]

| S9 Concentration (μl/plate) | Revertants/plate | | | | | |
|---|---|---|---|---|---|---|
| | 5 μg BaP | | 2.5 μg 2-AA | | 2.0 μg 2-AAF | |
| | Frozen | Lyophilized | Frozen | Lyophilized | Frozen | Lyophilized |
| 0 | 120 | 135 | 149 | 149 | 144 | 113 |
| 10 | 401 | 394 | 1817 | 1722 | 226 | 216 |
| 20 | 854 | 628 | 2258 | 2065 | 341 | 314 |
| 50 | 1105 | 1004 | 1480 | 1558 | 481 | 493 |
| 100 | 660 | 811 | 978 | 882 | 375 | 394 |
| 150 | 411 | 813 | 656 | 597 | 216 | 307 |

[a] BaP, Benzo(*a*)pyrene. 2-AA, 2-Aminoanthracene. 2-AAF, 2-Acetylaminofluorene.

TABLE A.4
Titration of Lyophilized and Frozen Aroclor-Induced Rat Liver S9 Homogenate against Benzo(*a*)Pyrene, 2-Acetylaminofluorene, and 2-Aminoanthracene Using TA98[a]

| S9 Concentration (μl/plate) | Revertants/plate | | | | | |
|---|---|---|---|---|---|---|
| | 5 μg BaP | | 2.5 μg 2-AA | | 2.0 μg 2-AAF | |
| | Frozen | Lyophilized | Frozen | Lyophilized | Frozen | Lyophilized |
| 0 | 73 | 54 | 73 | 77 | 76 | 75 |
| 10 | 340 | 344 | 1705 | 2022 | 1022 | 691 |
| 20 | 354 | 354 | 1880 | 1811 | 1040 | 737 |
| 50 | 509 | 395 | 1350 | 1420 | 958 | 775 |
| 100 | 232 | 232 | 549 | 702 | 553 | 557 |
| 150 | 139 | 309 | 442 | 618 | 435 | 347 |

[a] BaP, Benzo(*a*)pyrene. 2-AA, 2-Aminoanthracene. 2-AAF, 2-Acetylaminofluorene.

## Other S9 Problems

Beyond the basic problems discussed previously, new studies and applications of S9, especially as they apply to the area of *in vitro* mammalian cell culture, have identified aspects of S9 use that appear to be particularly relevant to study design and data analysis.

One of the first adverse properties of S9 homogenate identified was its potential toxicity for cultured cells. Cytotoxicity to cells is extremely

TABLE A.5
Titration of Lyophilized and Frozen Aroclor-Induced Rat Liver S9 Homogenates against Dimethylnitrosamine (DMN) Using TA1535 and TA100 in a Preincubation Assay[a]

| S9 Concentration (μl/plate) | Revertants/plate | | | |
|---|---|---|---|---|
| | TA1535 | | TA100 | |
| | Frozen | Lyophilized | Frozen | Lyophilized |
| 0 | 12 | 14 | 133 | 120 |
| 10 | 18 | 13 | 125 | 132 |
| 20 | 21 | 29 | 154 | 142 |
| 50 | 214 | 92 | 278 | 248 |
| 100 | 1242 | 831 | 2427 | 1879 |
| 150 | 1694 | 1604 | 2215 | 2169 |

[a] DMN = 100 μM.

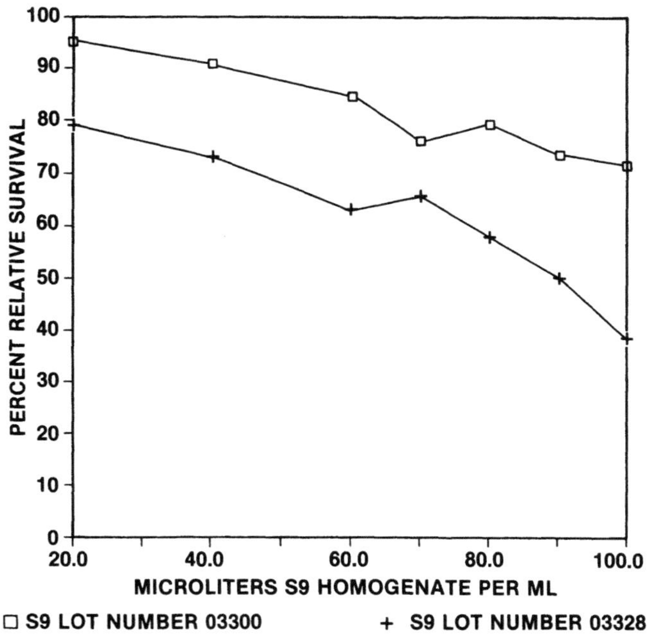


FIGURE A.3. A comparison of S9 homogenate cytotoxicity in a CHO cell clonal toxicity assay. Both batches showed equivalent activation of benzo(*a*)pyrene and 2-aminoanthracene to Ames mutagens. S9 lot number 03328 would be too toxic for cell culture use.

variable and cannot be predicted using biochemical measurements. Figure A.3 shows the $LC_{50}$ concentrations for two lots of S9 that are similar in biochemical measurements and in their bioactivation capacity in the Ames assay. These two batches represent examples of the extreme ranges of toxicity found among S9 batches. As a consequence of the variable toxicity among preparations, S9 quality control should include pretesting for cytotoxicity with one or more mammalian cell lines if it is intended for use in mammalian cell bioassays. An acceptance standard that seems to work is that S9 should not reduce the survival of CHO cells below 70% at 50 μl/ml and should not produce 100% lethality at 100 μl/ml.

Under certain conditions S9 has been demonstrated to undergo biochemical changes that result in the formation of mutagenic agents (possibly via lipid peroxidation). The mutagens appear to form as the result of enzymatic activities, since high temperatures will significantly reduce their formation.

Clearly, S9 and S9 mix are important reagents for genetic toxicologists. During the past 10 years these reagents have been used with little appreciation for their impact on test results and therefore only minimal quality control procedures have been developed. The potential for S9 variability is greater than for other reagents used, such as culture medium, buffer solutions, solvents, and various medium supplements, and efforts are needed to establish a set of acceptability criteria for S9 and S9 mix.

## REFERENCES


1. Malling, H. V.: Dimethylnitrosamine: Formation of mutagenic compounds by interaction with mouse liver microsomes. *Mutat. Res.* **13**:425–429, 1971.
2. Gabridge, M. G., and Legator, M. S.: A host-mediated microbial assay for the detection of mutagenic compounds. *Proc. Soc. Exp. Biol. Med.* **130**:831–834, 19699.
3. Arni, P., Mantel, T., Deparade, E., and Muller, D.: Intrasanguine host-mediated assay with *Salmonella typhimurium*. *Mutat. Res.* **45**:291–307, 1977.
4. Fahrig, R.: Development of host-mediated mutagenicity tests. I. Differential response of yeats cells injected into testes of rats and peritoneum of mice and rats to mutagens. *Mutat. Res.* **26**:29–36, 1974.
5. Ames, B. N., McCann, J., and Yamasaki, E.: Methods for detecting carcinogens and mutagens with the *Salmonella*/mammalian-microsome mutagenicity test. *Mutat. Res.* **31**:347–364, 1975.
6. Matheson, D., Brusick, D., and Carrano, R.: Comparison of the relative mutagenic activity for eight antineoplastic drugs in the Ames *Samonella*/microsome and TK+/− mouse lymphoma assays. *Drug Chem. Toxicol.* **1**:277–304, 1978.
7. Bartsch, H., Malaveille, C., and Montesano, R.: Human, rat and mouse liver-mediated mutagenicity of vinyl chloride in *S. typhimurium* strains. *Int. J. Cancer* **15**:429–437, 1975.
8. Brusick, D. J.: *In vitro* mutagenesis assays as predictors of chemical carcinogenesis in mammals. *Clin. Toxicol.* **10**:79–109, 1977.
9. Matushima, T., Sawamura, M., Hara, K., and Sugimura, T.: A safe substitute for polychlorinated biphenyls as an inducer of metabolic activation systems. In *In vitro Metabolic Activation in Mutagenesis Testing* (F. J. de Serres, J. R. Fouts, J. R. Bend, and R. M. Philpot, eds.), Elsevier/North-Holland Biomedical Press, Amsterdam, pp. 85–88, 1976.
10. Yoshikawa, K., Nohmi, T., Harada, R., Inokawa, Y., and Ishidate, Jr., M.: Differential mutagenicities of triaminobenzenes against *Salmonella typhimurium* TA98 in the presence of S9 fractions from polychlorinated biphenyls, phenobarbital and 3-methylcholanthrene-pretreated rats, hamsters and mice. *J. Toxicol. Sci.* **4**:317–326, 1979.
11. Prival, M. J., and Mitchell, V. D.: Influence of microsomal and cytosolic fractions from rat, mouse and hamster liver on the mutagenicity of dimethylnitrosamine in the *Salmonella* plate incorporation assay. *Cancer Res.* **41**:4361–4367, 1981.
12. Prival, M. J., and Mitchell, V. D.: Analysis of a method for testing azo dyes for mutagenic activity in *Salmonella typhimurium* in the presence of flavin mononucleotide and hamster liver S-9. *Mutat. Res.* **97**:103–116, 1982.
13. Dent, J. G., Schnell, S., Graichen, E., Allen, P., Abernathy, D., and Couch, D. B.: Stability of activating systems for *in vitro* mutagenesis assays: Enzyme activity and activating ability following long-term storage at −85°C. *Environ. Mutagen.* **3**:167–179, 1981.
14. Nohmi, T., Yoshikawa, K., Miyata, R., Nakamura, S., and Ishidate, Jr., M.: Duration of S9 activity in agar overlay. *Mutat. Res.* **91**:37–39, 1981.

## APPENDIX B: DOSE SELECTION FOR *IN VIVO* GENETIC ASSAYS

Selection of dose levels for *in vivo* studies is often one of the most critical factors in the entire operation. Dose levels that are suboptimal for the type of test being conducted could lead to erroneous conclusions when data are evaluated.

In general, genetic assays incorporate either acute or subchronic chemical administration. Exposure is commonly limited to a single or five daily administrations. Exceptions will be the mouse heritable translocation assay or the extended dosing modification for the dominant lethal assay, in which the chemical is administered continuously over the spermatogenic cycle of the male.

The common practice for dose determination is to perform a dose range experiment prior to the actual study. This usually consists of setting up 5–7 groups of animals for 4–8 animals per group. A series of doses is given to the groups on an acute basis using the anticipated route of exposure. If nothing is known about the toxicity of the test material, the maximum dose in the series might be 10 g/kg body weight, with lower doses in the series falling at natural logarithmic spaces or steps (e.g., 10, 3, 1, 0.3 g/kg, etc.).

Using either lethality or some other sentinel parameter, and $LD_{50}$ (lethal dose 50%) or $ED_{50}$ (effective dose 50%) level can be calculated by one of several methods.[1–3]

If the duration of exposure in the study is short and the animals are going to be evaluated shortly after exposure (6–48 hr), then a dose near the $LD_{50}$ or $ED_{50}$ level should be selected (e.g., $LD_{50}$).

This maximum is considered the maximum tolerated dose (MTD) and subjects the animals to levels of the chemical in excess of what might be expected in the environment, barring accidental release. It may be interpreted as the worst possible case. Typically, one or more lower doses are included in the study design, which permits evaluation at exposures near the anticipated use or environmental levels. If the data of a study using this dose selection scheme are negative, confidence can be had in extrapolating the results to larger populations.

If the data from the study show an effect, the shape of the response curve over several dose levels can be of importance in interpreting the likely effects near the anticipated use or environmental level. It is almost impossible to develop reliable interpretations from a study if the anticipated use or environmental level is used as the maximum dose, since all *in vivo* assays are characterized by limited sampling. Agents with weak to moderate activity could not be detected under such conditions.

When the dosing period is extended to 5 days or longer, it is very likely that the MTD will be smaller. This is not true for some relatively inocuous chemicals, but does apply to the majority of chemicals. The reasons are varied, but include:

1. The chemical may be incompletely cleared between administrations, thus increasing the total daily dose and leading to unanticipated toxicity.
2. Chemical interactions may lead to enzyme induction and possible alterations in metabolism with concomitant changes in toxicity.
3. Some chemicals produce delayed toxicity that is not observed in studies covering 1 or a few days, but is evident in more prolonged exposures.
4. Some chemicals may modify an animal's immune systems, leading to hypersensitivity or immune suppression. Either result can significantly alter the toxic effects of the chemical during extended dosing.

Therefore, before initiating studies such as the mouse heritable translocation assay, which may require 7–10 weeks of exposure, it will be necessary to conduct a long-term pilot study at several doses or reduce the acute MTD by some factor. For example, it is not uncommon to use $(\frac{1}{8}-\frac{1}{10})LD_{50}$ as the high dose in an extended exposure study design. Mid- and low-dose levels are usually selected as some fraction of the high dose (e.g., $\frac{1}{3}$ and $\frac{1}{10}$, respectively). This type of factoring is risky, since significant toxicity or lethality may result well into the study. A long-term pilot study, although costly, could prevent loss of the entire study and is the recommended method for dose selection.

Because so many of the *in vivo* assays for genetic endpoints involve mating sequences, it may be necessary to use an endpoint other than lethality to select the MTD. Animals with suppressed libido or nonviable sperm will not provide an accurate assessment of heritable genetic endpoints. Calculation of an $ED_{50}$ for a parameter such as fertility or neurologic effects might be appropriate for some chemicals. Regardless of the type of study, route of exposure, or duration of compound administration, it will be necessary to make regular (preferably twice daily) inspection for toxic signs.[4] Recording toxic signs is often an aid in the evaluation of the results, since it gives another parameter to the toxicity of the compound. Unusual findings at animal necropsy should also be recorded in the raw data.

## REFERENCES

1. Finney, D. J., *Statistical Method in Biological Assay,* Hafner Press, New York, 1969.
2. Litchfield, J. T., and Wilcoxson, F.: A simplified method of evaluating dose–effect experiments. *J. Pharmacol. Exp. Ther.* **96**:99–113, 1949.

3. Miller, L. C., and Tainter, M. L.: Estimation of the $ED_{50}$ and its error by means of logarithmic-probit graph paper. *Proc. Soc. Exp. Biol. Med.* **57**:621–64, 1944.
4. Campbell, P. E. S., and Richter, W.: An observation method estimating toxicity and drug actions in mice applied to 68 reference drugs. *Acta Pharmacol. Toxicol.* **24**:345–63, 1967.

# APPENDIX C: SELECTED REFERENCES AND REVIEWS OF GENETICS AND GENETIC TOXICOLOGY*

Dean, B. J. (ed.): Report of the UKEMS Subcommittee on Guidelines for Mutagenicity Testing Part I: Basic Test Battery; Minimal Criteria; Professional Standards; Interpretation; Selection of Supplementary Assays. UKEMS, Genetics Department, University College of Swansea, Swansea, Wales, Great Britain.

Dellarco, V. L. Voytek, F. E., and Hollaender, A. (eds.): *Aneuploidy: Etiology and Mechanisms,* Plenum Press, New York, 1985.

De Serres, F. J., and Hollaender, A. (eds.): *Chemical Mutagens: Principles and Methods for Their Detection,* Vols. 6–14, Plenum Press, New York, 1977–1986.

Douglas, J. F. (ed.): *Carcinogenesis and Mutagenesis Testing,* Humana Press, Clifton, New Jersey, 1984.

Fishbein, L., Flamm, W. G., and Falk, H. L. (eds.): *Chemical Mutagens,* Academic Press, New York, 1970.

Hollaender, A. (ed.): *Chemical Mutagens: Principles and Methods for Their Detection,* Vols. 1–4, Plenum Press, New York, 1971–1975.

Hollaender, A., and de Serres, F. J. (eds.): *Chemical Mutagens: Principles and Methods for Their Detection,* Vol. 5, Plenum Press, New York, 1976.

Kilbey, B. J., Legator, M., Nichols, W., and Ramel, C. (eds.): *Handbook of Mutagenicity Test Procedures,* Elsevier, Amsterdam, 1977.

National Academy of Sciences: *Identifying and Estimating the Genetic Impact of Chemical Environmental Mutagens,* National Academy Press, Washington, D. C., 1982.

Novitski, E.: *Human Genetics,* Macmillan, New York, 1977.

Office of Technology Assessment: *The Role of Genetic Testing in the Prevention of Occupational Disease,* Office of Technology Assessment, Washington, D. C., April 1983.

Scott, D., Bridges, B. A., and Sobels, F. H. (eds.): *Progress in Genetic Toxicology,* Elsevier/North-Holland, Amsterdam, 1977.

Strickberger, M. (ed.): *Genetics,* Macmillan, New York, 1968.

Sugimura, T., Kondo, S., and Takebe, H. (eds.); *Environmental Mutagens and Carcinogens,* University of Tokyo Press, Tokyo, and Allan R. Liss, New York, 1982.

* Additional information is available through the Environmental Mutagen Information Center (EMIC) at the Oak Ridge National Laboratories. This service maintains many manuscripts and publications on environmental mutagens (excluding radiation alone) that have been printed since 1969 and can answer bibliographic and reference questions.

The service is directed by John Wasson at: Environmental Mutagen Information Center, Oak Ridge National Laboratory, Building 2001, Mailstop 50, Post Office Box X, Oak Ridge, Tennessee 37831-6050. Telephone: (615) 574-7871.

Vogel, F., and Röhrborn, G. (eds.): *Chemical Mutagenesis in Mammals and Man*, Springer-Verlag, New York, 1970.
Watson, J.: *Molecular Biology of the Gene*, W. A. Benjamin, New York, 1970.
Woodhead, A. D., Shellabarger, C. J., Pond, V., and Hollaender, A. (eds.): *Assessment of Risk from Low-Level Exposure to Radiation and Chemicals*, Plenum Press, New York, 1985.

# Index

GPSR Compliance

*The European Union's (EU) General Product Safety Regulation (GPSR) is a set of rules that requires consumer products to be safe and our obligations to ensure this.*

*If you have any concerns about our products, you can contact us on ProductSafety@springernature.com*

In case Publisher is established outside the EU, the EU authorized representative is:

Springer Nature Customer Service Center GmbH
Europaplatz 3
69115 Heidelberg, Germany

**Batch number: 10620143**

Printed by Printforce, the Netherlands